Lecture Notes in Mathematics

Edited by A. Dold and B. Eckmann

Subseries: Institut de Mathématique,
Université de Strasbourg

Adviser: P. A. Meyer

1193

Geometrical and Statistical Aspects of Probability in Banach Spaces

Actes des Journées SMF de Calcul des Probabilités dans les Espaces de Banach, organisées à Strasbourg les 19 et 20 juin 1985

Edited by X. Fernique, B. Heinkel,
M. B. Marcus and P. A. Meyer

Springer-Verlag
Berlin Heidelberg New York Tokyo

Editors

Xavier Fernique
Bernard Heinkel
Paul-André Meyer
Institut de Recherche Mathématique Avancée
7 rue René Descartes
67084 Strasbourg Cedex, France

Michael B. Marcus
Department of Mathematics
Texas A & M University
College Station, Texas 77843, USA

Mathematics Subject Classification (1980): 46B20, 60B05, 60B10, 60B12, 60F05, 60F15, 60F17, 60G15, 62D05, 62E20

ISBN 3-540-16487-1 Springer-Verlag Berlin Heidelberg New York Tokyo
ISBN 0-387-16487-1 Springer-Verlag New York Heidelberg Berlin Tokyo

Printing and binding: Beltz Offsetdruck, Hemsbach/Bergstr.
2146/3140-543210

Préface

Le calcul des probabilités dans les espaces de Banach est actuellement un sujet en plein essor auquel des rencontres internationales sont consacrées régulièrement depuis une dizaine d'années. Les 19 et 20 juin 1985, une trentaine de spécialistes de ce sujet se sont réunis à Strasbourg sous le patronage de la Société Mathématique de France, pour faire le point des développements les plus récents, notamment en matière de fonctions aléatoires gaussiennes, de processus empiriques et de théorèmes limites pour des variables aléatoires à valeurs dans un espace de Banach. Les principaux exposés de ces deux journées ont été rédigés par leurs auteurs, ce qui a permis de composer ces Actes que la Société Springer a eu l'amabilité d'accueillir dans sa collection Lecture Notes in Mathematics.

Ces deux journées ont été assombries par la disparition, le 7 juin 1985, d'Antoine Ehrhard qui était l'un des plus brillants représentants de la jeune génération de probabilistes. Nous avons ressenti cruellement son absence, celle du mathématicien bien sûr, mais surtout celle de l'homme de coeur sensible et attachant qu'il était.

Les éditeurs

Table of Contents

A BRIEF SURVEY OF ANTOINE EHRHARD'S SCIENTIFIC WORK

Christer BORELL
Dept. of Mathematics
Chalmers University of Technology
Göteborg, Sweden

For a couple of years Antoine Ehrhard gave us pleasure with a series of brillant ideas on Gaussian measures and convexity. The loss of him is the loss of a very seriously working young mathematician as well as the loss of a very good friend. For a complete list of Antoine Ehrhard's publications, see at the end of this survey.

As a background to Ehrhard's scientific work it is appropriate to recall the Laplace-Beltrami operator and its relations to isoperimetry. Needless to say, this fascinating area is far from completed and, indeed, it seems very hard to unify since it is rooted in so many branches of pure and applied mathematics. In particular, like the Laplace operator $\Delta = \nabla^2$ in $\mathbb{R}^n$, the so-called number operator $L = -\Delta + x.\nabla$ in $\mathbb{R}^n$ merits its own study. This central point underlines most of Ehrhard's papers.

To master the number operator and isoperimetry Ehrhard first introduced the so-called k-dimensional Gaussian symmetrizations and he developed a streamlined machinery of general interest [MS]. For brevity, we only recall the definition of n-dimensional Gaussian symmetrizations.

Suppose μ is the canonical Gaussian measure in $\mathbb{R}^n$ i.e.

$$\mu(dx) = e^{-|x|^2/2}\, dx/\sqrt{2\pi}^{\,n}$$

and let $h \in \mathbb{R}^n$ be a fixed unit vector. Set $\tilde{h} = <.,h>_{\mathbb{R}^n}$. Then to any $f \in L_0(\mu)$ there exists a unique non-decreasing $g \in L_0(\tilde{h}(\mu))$ such that $f(\mu) = g(\tilde{h}(\mu))$. The function $f^* = g \circ \tilde{h}$ is called the n-dimensional Gaussian symmetrization of f in the direction of h. Here, if f is the indicator function of a set, f^* becomes the indicator function of an affine half-space.

The number operator L is related to Gaussian Dirichlet integrals as follows :

$$\int f(Lf)\,d\mu = \int |\nabla f|^2 d\mu \quad , \; f \in C_0^\infty(\mathbb{R}^n) \quad .$$

Ehrhard's perhaps most central result states that the integral

$$\int F(|\nabla f|)\,d\mu$$

decreases, in the weak sense, under Gaussian symmetrizations of f for every non-decreasing convex function $F : [0,+\infty[\; \to \mathbb{R}$ [ASE]. The familiar isoperimetric inequalities for torsional rigidity, principal frequency, and Newtonian capacity thereby get their Gaussian counterparts now with affine half-spaces as extremals [ASE].

The same source of ideas also led Ehrhard to a very neat proof of the Gross logarithmic Sobolev inequality [LN] and to an inequality of the Poincaré type [ASE].

Under the leadership of Professor Xavier Fernique, Antoine Ehrhard very early became familiar with the Banach space aspect of stochastic processes, which has been of greatest significance to Ehrhard's maturity as a mathematician. For an important joint publication, see [CR]. Another result stemming from this background is the following remarkable inequality :

(1) $\Phi^{-1}(\mu(\theta A+(1-\theta)B)) \geq \theta\Phi^{-1}(\mu(A))+(1-\theta)\Phi^{-1}(\mu(B))$, $0<\theta<1$, A,B convex,

where

$$\Phi(a) = \int_{-\infty}^{a} e^{-\lambda^2/2}\, d\lambda / \sqrt{2\pi}$$

[MS]. In particular, if A is a convex body in $\mathbb{R}^n$ and if u is the solution of the heat equation in $(\operatorname{int} A) \times \mathbb{R}_+$ satisfying $u=0$ on $(\partial A) \times \mathbb{R}_+$ and $u=1$ on $A \times \{0\}$, then the function $\Phi^{-1}(u(\cdot,t))$ is concave for every $t>0$ [MS].

Finally, in his last paper [AIP] Ehrhard investigated the case of equality in (1) as well as in several other inequalities for Gaussian measures. Thus, for example, if $A,B \subsetneq \mathbb{R}^n$ are non-empty convex domains, Ehrhard proved that equality occurs in (1) if and only if either $A=B$ or A and B are parallel affine half-spaces. The arguments leading to this very definite result are extremely penetrating and mixed with youthful enthusiasm and conviction.

The loss of Antoine Ehrhard is an irreplaceable loss to the area he so successfully invented. The scientific progress will now proceed much slower and with much less substance, too. But for ever, we will remember an artist ; an artist painting with convex bodies and the Gaussian law.

PUBLICATIONS

[CR] : (en collaboration avec X. Fernique). Fonctions aléatoires stables irrégulières. C.R. Acad. Sc. Paris, t.292, Série I (1981), 999-1001.

Régularité des fonctions aléatoires stables.
Thèse de 3ème cycle (1982), 9-43.

Lois stables et propriété de Slépian.
Ann. Sc. de l'Université de Clermont 71 (1982), 81-94.

Sur la densité du maximum d'une fonction aléatoire gaussienne.
Séminaire de Probabilités XVI, 1980/81, Lecture Notes in Math. 920, 581-601.

Une démonstration de l'inégalité de Borell.
Ann. Sc. de l'Université de Clermont 69 (1981), 165-184.

[MS] : Symétrisation dans l'espace de Gauss.
Math. Scand. 53 (1983), 281-301.

Un principe de symétrisation dans les espaces de Gauss.
Probability in Banach spaces IV -Oberwolfach 1982- .
Lecture Notes in Math. 990, 92-101.

[ASE] : Inégalités isopérimétriques et intégrales de Dirichlet gaussiennes.
Ann. Scient. Ec. Norm. Sup., 4è série, t. 17 (1984), 317-332.

[IN] : Sur l'inégalité de Sobolev logarithmique de Gross.
Séminaire de Probabilités XVIII, 1982/83,
Lecture Notes in Math. 1059, 194-196.

[AIP] : Eléments extrémaux pour les inégalités de Brunn-Minkowski gaussiennes.
Annales de l'Institut Henri Poincaré. Vol. 22 n° 1 (1986) , 149-168.

Sur la densité du maximum d'un processus gaussien.
Thèse d'Etat (1985), 87-104.

Thèse de 3ème cycle (soutenue à Strasbourg le 12.02.1982) :
Fonctions aléatoires stables. Densité du maximum d'une fonction aléatoire gaussienne.
Publication de l'IRMA de Strasbourg, n° 156.

Thèse d'Etat (soutenue à Strasbourg le 24.05.1985) :
Convexité des mesures gaussiennes.
Publication de l'IRMA de Strasbourg, n° 273.

INVARIANCE PRINCIPLES FOR THE EMPIRICAL MEASURE OF A MIXING SEQUENCE AND FOR THE LOCAL TIME OF MARKOV PROCESSES.

P. DOUKHAN*, J.R. LEON**

ABSTRACT.

We show an invariance principle for the empirical measure of a stationary strongly mixing sequence indexed by the unit ball of some Sobolev space H_s. We also obtain invariance principle and law of iterated logarithm for the local time of Markov processes indexed by H_s.

We note that the regularity condition $s > d/2$ in the first framework for random variables with values in a compact riemannian manifold E becomes $s > d/2-1$ in the continuous case of the brownian motion on E.

*Université Paris-Sud
U.A. CNRS 743 "Statistique Appliquée"
Mathématique, Bât. 425
91405 ORSAY (France)

**Universidad Central de Venezuela
Facultad de Ciencias
Departamentado de Matematicas
Apartado Postal n° 21201
CARACAS (Vénézuela).

1. INTRODUCTION

This work is divided in two parts. The first one is devoted to investigate a rate of convergence in the weak invariance principle for the empirical process X_n of a strictly stationary strongly mixing sequence $\{\xi_k \ ; \ k=0,1,\dots\}$ valued in a metric space E indexed by a compact class F of $L^2(E,\mu)$ of uniformly bounded functions satisfying an entropy condition : $X_n(f) = \frac{1}{\sqrt{n}} \sum_{k=1}^{n} [f(\xi_k)-Ef(\xi_k)]$, $f \in F$.

The typical case is obtained for a d-dimensional riemannian compact manifold E with F unit ball of the Sobolev space H_s of the manifold (see Giné [14]) ; a result can be shown only if $s > d/2$. In this discrete case we expose some of the results of [10] made in collaboration with Frédéric Portal. Rates of convergence essentially depend on the entropy condition for F .

The second part of this paper studies the asymptotic behaviour of $Z_n(f) = \frac{1}{\sqrt{n}} \int_0^n f(X_u)\, du$, $f \in L^2(\mu)$. Here $\{X_t : t > 0\}$ is a continuous parameter recurrent ergodic stationary Markov process with values in a compact riemannian manifold E or in $\mathbb{R}^d$; μ denotes the invariant measure of the process. We first give an invariance principle in a general framework. We also study the case of the brownian motion on E ; we give an invariance principle and a L.I.L. uniform on the classe F , unit ball of H_s for $s > d/2-1$. We also study the case of diffusions on $\mathbb{R}^d$. Finally we discretize the Z_n process by :

$$Z_{n,\Delta}(f) = \frac{1}{\sqrt{n}} \sum_{0<k\Delta\leq n} f(X_{k\Delta}) \quad , \quad f \in F.$$

We give condition of uniform convergence of this discretization.

Our result can be compared to those of [7], which works with the non-uniform case of one dimensional diffusions. Our speed of discretization is quite lower but our result is uniform on F. The process $Z_n(f)$ has been studied for the case of E by Baxter and Brosamler [4]; they prove non uniform central limit theorem and L.I.L. They use, as we do, mixing techniques. Battacharya ([3]) extends their results to a non compact case with martingale techniques. A first uniform result, based on a martingale approach, is given by Bolthausen [5] for the case of brownian motion on the d-dimensional torus ; his results are extended here for a general riemannian compact manifold. The non uniform problem is also considered by D. Florens [13] for the case of one dimensional diffusions.

We use, in this paper, mixing and Hilbert space techniques which are by-products of [10] for example. The Markov case satisfies a mixing assumption as it is shown in Rosenblatt [21] . The mixing notion used here is strong mixing defined as follows $\{x_\nu ; \nu \geq 0\}$ is said to be strongly mixing, with mixing coefficient α_ν if $\alpha_\nu \underset{\nu\to\infty}{\longrightarrow} 0$ where,

$$\alpha_\nu = \operatorname{Sup}\{|\mathbb{P}(A\cap B) - \mathbb{P}(A)\,\mathbb{P}(B)| \; ; A \in F_0^\theta \,, B \in F_{\theta+\nu}^\infty, \theta \geq 0\}$$

and F_s^t is the σ-field generated by $\{x_\nu ; s \leq \nu \leq t\}$.

We expose here invariance principles results for discrete and continuous parameter processes to compare them. In the compact riemannian manifold case we have seen that regularity conditions $s > d/2$ for discrete parameter becomes $s > d/2-1$ for continuous parameter.

2. INVARIANCE PRINCIPLE FOR THE EMPIRICAL MEASURE OF A MIXING SEQUENCE OF RANDOM VARIABLES.

In this section we expose some results of [10] made in collaboration with Frédéric Portal ; a complete version of this work will appear elsewhere with proofs.

Let $(\xi_k)_{k\geq 0}$ a strictly stationary sequence of strongly mixing random variables with values in a polish measured space $(E,\mathcal{B}(E),\mu)$ where $\mathcal{B}(E)$ is the Borel σ-field of E and μ a non negative σ-finite measure. Let F a finite entropy compact subset of $L^2(E,\mu)$, we define :

$$X_n(f) = \frac{1}{\sqrt{n}} \sum_{k=1}^{n} [f(\xi_k)-Ef(\xi_k)] \,, \qquad f \in F$$

The class F is supposed uniformly bounded, the law of ξ_0 has a bounded density with respect to μ , and there is some $\sigma\in]0,1/3]$ such that $\sum_{k=0}^{\infty} \alpha_k^\sigma < \infty$ (here $\{\alpha_k\}$ denote the mixing coefficients of the sequence $\{\xi_k\}$). The process X_n is $C(F)$ valued, where $(C(F), \|\cdot\|_\infty)$ is the space of continuous functions on the compact set F equipped with uniform norm.

We give an estimate of Prohorov distance of X_n to the centered gaussian process Y with covariance defined for $f,g \in F$ by :

$$EY(f)\,Y(g) = Ef(\xi_0)g(\xi_0) - \bar{f}\bar{g} + \sum_{k=1}^{\infty} E\,\{(f(\xi_0)-\bar{f})\,(g(\xi_k)-\bar{g}) + (f(\xi_k)-\bar{f})(g(\xi_0)-\bar{g})\}$$

here $\bar{f} = Ef(\xi_0)$ and $\bar{g} = Eg(\xi_0)$.

A reconstruction of the process Y gives a weak invariance principle with rate of convergence.

The method is based on estimations of central limit theorem rates in Prohorov metric convergence given in [11] for the finite repartitions of the process X_n depending on the dimension of repartitions and, from another hand, on estimations of the oscillations of the X_n process based on,

<u>*THEOREM 1*</u>. *Let* $F_\delta = \{f \in F\text{-}F \;;\; \int f^2\, d\mu < \delta^2\}$ *, suppose that there is a* $\sigma \in]0,1/3]$ *satisfying* $\sum_{k=0}^{\infty} \alpha_k^\sigma < \infty$*, then there is a constant only depending on the* $\{\xi_k\}$ *sequence such that :* $E\,\sup\,\{X_n^2(f) \;;\; f \in F_\delta\} \leq C\,\{\delta^2 U + B_\delta V\}$.

Let $\{f_k \;;\; k \in \mathbb{K}\}$ *an orthonormal denumerable basis of* $L^2(E,\mu)$ *, the values of* U,V,B_δ *are defined for* N *sufficiently big finite subset of* $\mathbb{K}$ *by :*

$$U = \left[\int_E \left(\sum_{k\in N} f_k^2(x)\right)^{1/(1-\sigma)} \mu(dx)\right]^{1-\sigma} \quad \left(\text{resp. } U = \sum_{k\in N} \|f_k\|_\infty^2\right) ,$$

$$V = \left[\int_E \left(\sum_{k\notin N} \theta_k^{-1}\, f_k^2(x)\right)^{1/(1-\sigma)} \mu(dx)\right]^{1-\sigma} \quad \left(\text{resp. } V = \sum_{k\notin N} \theta_k^{-1}\, \|f_k\|_\infty^2\right) ,$$

$$B_\delta = \operatorname{Sup}\left\{\sum_{k\notin N} \theta_k (f,f_k)^2 \;;\; f \in F_\delta\right\} ,$$

the sequence $\{\theta_k \;;\; k \in \mathbb{K}\}$ *is positive and satisfies* $\operatorname{Sup}\left\{\sum_{k\in \mathbb{K}} \theta_k (f,f_k)^2 \;;\; f \in F\right\} < \infty$.

<u>Remark</u> : The same estimate is valid for the gaussian process Y (see [10]) .

2.1. <u>Uniformity test on a d -dimensional compact riemannian manifold E</u> .

Here μ is the uniform measure on E and the law ν of ξ_0 satisfies : $\|\frac{d\nu}{d\mu}\|_\infty \leq a < \infty$. We extend results of Giné [14] in a mixing framework .

The test of Giné rejects hypothesis "$\nu = \mu$" for big values of $T_n^{(s)}$:

$$T_n^{(s)}(\omega) = n \left\| \sum_{k=0}^{\infty} (a_k \, \sigma_k^{s/2} \, \pi_k)(\nu_n(\omega) - \nu) \right\|_{-s}^2 .$$

The sequence $\{\sigma_k \ ; \ k=0,1,\ldots\}$ is the ordered sequence of eigenvalues of the Laplace Beltrami operator Δ with eigenspace E_k and π_k is the orthogonal projection $H_{-s} \to E_k$, H_s being the index s Sobolev space of the manifold $E (s \in \mathbb{R})$ whose norm is $\|f\|_s = \left(\sum_{k=0}^{\infty} \sigma_k^s \sum_{f_j \in E_k} (f,f_j)^2 \right)^{1/2}$. We write ν_n the empirical random measure : $\nu_n = \frac{1}{n} \sum_{k=1}^{n} \delta_{\xi_k}$. The sequence $\{a_k \ ; \ k = 0,1,\ldots\}$ satisfies $\operatorname{Sup} \{ |a_k \, \sigma_k^{s/2}| \ , \ k=0,1,\ldots \} < \infty$.

For this we consider the X_n process which can be written :

$$X_n(f) = \sqrt{n} \int_E f \, d(\nu_n - \nu) .$$

THEOREM 2. If there exist $0 < b < 1/4$, $0 < \sigma \leq 1/5$ *satisfying* $\sum_{n=0}^{\infty} n^2 \alpha_n^{\sigma} < \infty$ *and* $\alpha_n^{\sigma} = 0 \, (n^{-2/(3b)+8/15})$ *for* $n \to \infty$, *we can construct an identically distributed sequence of gaussian processes* $\{Y_n \ ; \ n=1,2,\ldots\}$ *with the law of* Y *such that :*

$$\mathbb{P} \left(\operatorname{Sup} \{ |X_n(f) - Y_n(f)| \ ; \ f \in B_s \} \geq \rho_n \right) \leq \rho_n$$

for $s > d/2$, *here* $\rho_n = 0 \left((\ell n(\ell n \, n))^{s/d} \, (\ell n \, n)^{(d-2s)/(3d)} \right)$ *for* $n \to \infty$ *and* B_s *denotes the closed unit ball of* H_s .

We use this result to show that, if "$\nu = \mu$" then the limit law of $T_n^{(s)}$ is a sum of dependent χ^2 random variables ; from another hand, for "$\nu \neq \mu$" satisfying $\pi_k(\nu) \neq 0$ for some k such that $a_k \neq 0$ then we show that the limit law of $n^{-1/2} \, (T_n^{(s)} - nC)$ is gaussian for some constant C .

So that the test is consistent.

In the case when $E = S^1$ is the unit circle equiped with a set of Hölder continuous functions with index $s > 1/2$ we obtain a more precise statement.

2.2. Invariance principle for the empirical measure on the real line.

Here $E = \mathbb{R}^1$, μ is Lebesgue measure on $\mathbb{R}^1$ and $F = \{ f \in L^2(\mathbb{R}^1) \ ; \ \sum_{k=0}^{\infty} (k+1)^s \, (f,h_k)^2 \leq 1 \}$ where $\{h_k\}$ the sequence of normalised Hermite functions (c.f. [23]) .

THEOREM 3. [9] . *Under the same mixing assumptions than in theorem 2, we obtain* $\rho_n = 0\ (\sqrt{\ell n\ (\ell n\ n)}\ (\ell n\ n)^{(5-6s)/18})$ *for* $n \to \infty$ *if* $s > 5/6$.

Remark :

. The main interest of this result is to escape from a compact context.

. In [10] and [9] invariance principles for non parametric estimators are obtained with the same methods. Kernel and projection estimators of density and regression function are considered. In this case the class F varies with the index n .

3. INVARIANCE PRINCIPLE AND L.I.L. FOR THE LOCAL TIME OF CONTINUOUS PARAMETER MARKOV PROCESSES.

We write $\{X_t\ ;\ t \geq 0\}$ a continuous parameter Markov process with values in a complete separable metric space E . This process is supposed to be homogeneous and stationary with marginal law μ . Moreover, its infinitesimal generator L is an unbounded non-negative linear operator, whose domain, $D(L)$ is dense in $L^2(\mu)$, and satisfies :

(i) L is self-adjoint and onto.

(ii) the spectrum of L is discrete.

(iii) 0 is a single eigenvalue of L associated to the constant eigenfunction 1 .

Under those hypotheses, the Hilbert space $L^2(\mu)$ is separable ; we write $\|\cdot\|$ (resp. $(.,.)$) its norm (resp. scalar product) and $\{e_m\ ;\ m = 0,1,2,\dots\}$ is an orthonormal basis of $L^2(\mu)$ such that :

$$Lf = \sum_{m=0}^{\infty} \lambda_m (f,e_m) e_m \quad \text{for } f \in D(L), \text{ here } \lambda_0 = 0 < \lambda_1 \leq \lambda_2 \leq \dots \text{ and } e_0 = 1 .$$

Thus the spectrum of L is $\{\lambda_m\ ;\ m \geq 0\}$. The semi-group P_t associated to the Markov process, defined by $(P_t f,g) = Ef(X_t)\ g(X_0)$, satisfies :

$$P_t f = \sum_{m=0}^{\infty} e^{-\lambda_m t} (f,e_m)\ e_m \quad , \quad f \in L^2(\mu)\ ,\ t \geq 0 \quad .$$

Condition (iii) implies ergodicity of the process $\{X_t\ ;\ t \geq 0\}$ (cf. Battacharya [3], proposition 2.2) . The operator P_t is a contraction verifying

$\|P_t f\| \leq e^{-\lambda_1 t} \|f\|$ for $f \in L^2(\mu)$, $(f,1) = 0$; thus the process $\{X_t \; ; \; t \geq 0\}$ is strongly mixing with $\alpha_t \leq c\, e^{-\lambda_1 t}$ for a $c > 0$ (Rosenblatt [21]) .

The Green operator G is defined on $L^2(\mu)$ by $Gf = \int_0^\infty P_t f \, dt$ if $f \in 1^\perp$ and $G1 = 0$ thus :

$$Gf = \sum_{m=1}^{\infty} \lambda_m^{-1} (f,e_m)\, e_m \; , \; f \in 1^\perp \quad .$$

This operator is continuous on $L^2(\mu)$ with range $D(L) \cap 1^\perp$ and verifies $LG = I - S$ where I is the identity operator on $L^2(\mu)$ and S is the orthogonal projection on e_0 $(Sf = (f,e_0)e_0)$.

We consider the Hilbert space H_s for $s \geq 0$ with norm (resp. scalar product) $\|.\|_s$ (resp. $(.,.)_s$) defined by :

$$H_s = \left\{ f \in L^2(\mu) \; ; \; \sum_{m=1}^{\infty} \lambda_m^s (f,e_m)^2 < \infty \right\} \; ; \; \|f\|_s = \left[\sum_{m=1}^{\infty} \lambda_m^s (f,e_m)^2 \right]^{1/2} \quad .$$

The operators L^r and G^r are formally described by Seeley [20] for $r \in \mathbb{C}$. Note that H_s is the range of $G^{s/2}$ and the domain of $L^{s/2}$. We also write H_{-s} the dual space of $(H_s, \|.\|)$, it is an Hilbert space with the norm $\|.\|_{-s}$:

$$\|T\|_{-s} = \text{Sup}\,\{|T(f)| \; ; \; f \in H_s \; , \; \|f\|_s = 1\} = \left\{ \sum_{m=1}^{\infty} \lambda_m^{-s} (T(e_m))^2 \right\}^{1/2}$$

We write B_s the unit closed ball of H_s .

The aim of this work is the study of asymptotic behaviour of the functionnal :

$$Z_n(f) = \frac{1}{\sqrt{n}} \int_0^n f(X_u)\, du \quad , \quad f \in 1^\perp \cap L^2(\mu)$$

It is defined on $L^2(\mu)$ because $E\,\|Z_n(f)\|^2 \leq n\,\|f\|^2$. We consider Z_n as an H_{-s} valued random variable for $s > 0$ verifying :

$$(*) \qquad \sum_{m=1}^{\infty} \lambda_m^{-(1+s)} < \infty \quad .$$

Indeed, $E\,\|Z_n\|_{-s}^2 = E \sum_{m=1}^{\infty} \lambda_m^{-s} \left(\frac{1}{\sqrt{n}} \int_0^n e_m(X_u)\, du \right)^2$

$$E\|Z_n\|^2_{-s} = 2\sum_{m=1}^{\infty} \lambda_m^{-s} \int_0^n (1-\frac{u}{n})\,(P_u\, e_m, e_m)\, du$$

$$E\|Z_n\|^2_{-s} = 2\sum_{m=1}^{\infty} \lambda_m^{-s} \int_0^n (1-\frac{u}{n})\, e^{-\lambda_m u}\, du$$

$$E\|Z_n\|^2_{-s} = 2\sum_{m=1}^{\infty} \lambda_m^{-(1+s)} - \frac{2}{n}\sum_{m=1}^{\infty} \lambda_m^{-(2+s)}\,(1-e^{-n\lambda_m})$$

This calculus is analogous to those of Baxter and Brosamler ([4] , theorem (4.3)).

We now define a gaussian random vector Z on H_{-s} . For this we assume the following technical hypothesis:

(H) $\left\{\right.$ There is a uniform random variable on the interval $[0,1]$ defined on the same probability space $(\Omega, A, \mathbb{P})$ that $\{X_t\ ;\ t \geq 0\}$ and independent of it.

We consider an i.i.d. sequence $(\xi_m)_{m>0}$ of gaussian centered random variables with variance 1 and we define, for s satisfying (*) :

$$Z(f) = \sqrt{2}\sum_{m=1}^{\infty} \lambda_m^{-1/2}\,(f, e_m)\,\xi_m\ , \qquad f \in B_s$$

Note that $E\|Z\|^2_{-s} = 2\sum_{m=1}^{\infty} \lambda_m^{-(1+s)}$, thus $Z \in H_{-s}$ a.s.

An abstract construction of Z can be made using the operator $T = 2\, G^{1+s}$ whose trace is $2\sum_{m=1}^{\infty} \lambda_m^{-(1+s)} < \infty$. We define a gaussian measure M on H_{-s} , under hypothesis (*), satisfying :

$$\int_{H_{-s}} (z,T_1)_{-s}\,(z,T_2)_{-s}\, M(dz) = (\Gamma\, T_1,T_2)_{-s}\ ,\ T_1,T_2\ \ H_{-s}\ .$$

The law of Z is M .

We also note that measurability condition of [3] is satisfied here because of the separability of E .

The random variables Z_n and Z , H_{-s} - valued, can be considered as $C(B_s)$ valued processes. Here we note $C(B_s)$ the space of continuous functions defined on the compact subset B_s of $L^2(\mu)$ and real valued equiped with uniform norm .

<u>*THEOREM 4*</u>. *The sequence* $\{Z_n\ ;\ n \geq 1\}$ *converges in distribution to* Z *in* $C(B_s)$ *, under the hypotheses (i), (ii), (iii) and (*)* .

Proof. Under the former hypotheses, Battacharya shows convergence of finite repartitions ([3], remark 2.1.1.). The tightness of the sequence $\{Z_n \ ; \ n \geq 1\}$ will result from flattly concentrated property using De Acosta's method [1] .

Let $F_m \subset H_{-s}$ be the m-dimensional space defined by :

$$F_m = \{T \in H_{-s} \ ; \ T\ e_k = 0 \quad \text{for} \quad k = 0 \quad \text{and} \quad k > m\} \ .$$

We note F_m^ε the ε-vicinity of F_m in H_{-s} and :

$$M_m^{+} = \left\{T \in H_{-s} \ ; \ T\ e_o = 0 \ , \ \sum_{k=m+1}^{\infty} \lambda_k^{-s} (T\ e_k)^2 < \varepsilon^2\right\}$$

Note that $M_m^\varepsilon \subset F_m^\varepsilon$ and $E\ Z_n^2\ (e_k) < 2/\lambda_k$, so Bienaymé-Tchebičev inequality implies :

$$\mathbb{P}\ (Z_n \notin F_m^\varepsilon) \geq \mathbb{P}\ (Z_n \notin M_m^\varepsilon)$$

$$\mathbb{P}\ (Z_n \notin E_m^\varepsilon) \geq 1 - \mathbb{P}\left(\sum_{k=m+1}^{\infty} \lambda_k^{-s}\ Z_n^2\ (e_k) \geq \varepsilon^2\right)$$

$$\mathbb{P}\ (Z_n \notin F_m^\varepsilon) \geq 1 - \varepsilon^{-2} \sum_{k=m+1}^{\infty} \lambda_k^{-s}\ E\ Z_n^2\ (e_k)$$

$$\mathbb{P}\ (Z_n \notin F_m^\varepsilon) \geq 1 - 2\varepsilon^{-2} \sum_{k=m+1}^{\infty} \lambda_k^{-(1+s)}$$

Thus the sequence Z_n is flattly concentrated ; theorem 4 follows.

COROLLARY 5. The sequence of real random variables $\|Z_n\|_{-s}^2$ *converges in distribution to* $\|Z\|_{-s}^2$ *, infinite sum of weighted* χ^2 *random variables.*

Remark : Using the direct construction of Z, note that : $\|Z\|_{-s}^2 = 2 \sum_{m=1}^{\infty} \lambda_m^{-(1+s)} \xi_m^2$, $(\xi_m)_{m>0}$ being an i.i.d. sequence of normal random variables.

In view to investigate iterated logarithm behaviour of Z_n, we now make a direct construction of the brownian process Z_t with base M, law of Z . We use the Levy's construction ([16], 1.5, p. 19). Let

$\{\chi_{n,k} \ ; \ n = 0,1,\ldots, \ k \text{ odd}, \ k = 1,\ldots,2^n-1\}$ the Haar basis of $L^2\ [0,1]$ defined by : $\chi_{n,k}(t) = 2^{-n/2}$ for $k \leq 2^n t \leq k+1$, $= 0$ else, and $\tilde{Z}_{n,k}$ an i.i.d. array of realizations of Z defined on $(\Omega, A, \mathbb{P})$, we write $\tilde{Z}_N(t) = \sum_{n=0}^{N} \sum_k \tilde{Z}_{n,k} \int_0^t \chi_{n,k}(u)\ du$. This serie converges normally a.s. Note, for this, that :

$E_N = \text{Sup}\ \{\ \|\tilde{Z}_N(t)-\tilde{Z}_{N-1}(t)\|_{-s}\ ;\ 0 \leq t \leq 1\} \leq 2^{-(N+1)/2}\ \text{Max}\ \{\ \|\tilde{Z}_{N,k}\|_{-s}\ ;\ k \text{ odd} < 2^N\}$

From another hand Fernique ([12], theorem 1.3.2.) shows that there is an $a > 0$ such that : $A = E \exp(-\|Z\|^2_{-s}/a^2) < \infty$. Thus $\mathbb{P}(E_N \geq a\ 2^{-(N+1)/2} \sqrt{\ell n(2^{2N})}) \leq A\ 2^{-N}$.

Borel-Cantelli lemma implies then the continuity of the limit $\tilde{Z}(t)$ of the sequence $Z_N(t)$. The limiting covariance is computed as :

$$E\ (\tilde{Z}(u),U)_{-s} = (\tilde{Z}(V),V)_{-s} = \sum_{n=0}^{\infty} \sum_{\substack{k \text{ odd} \\ 0<k<2^n}} E(Z,U)_{-s}\ (Z,V)_{-s} \int_0^{u \wedge v} \chi_{n,k}(w)\ dw$$

$$E\ (\tilde{Z}(u),U)_{-s}\ (\tilde{Z}\ (v),V)_{-s} = u \wedge v\ E(Z,U)_{-s}\ (Z,V)_{-s}$$

Let $\{Z(n,t)\ ;\ 0 \leq t \leq 1\}$ an i.i.d. sequence of continuous realizations of $\tilde{Z}(t)$, we set $Z(t) = \sum_{k=1}^{[t]} Z(k,1) + Z\ ([t],\ t-[t])$, where $[t]$ denotes the integer part of t. The process $\{Z(t)\ ;\ t \geq 0\}$ is the brownian process with basis M ; it belongs to $C\ (\mathbb{R}^+, H_{-s})$ under condition (*).

PROPOSITION 6. The a.s. cluster set of $Z(t)\ (2\ t\ \ell n[\ell n\ t])^{-1/2}$ *for* $t \to \infty$ *is the compact subset* $\sqrt{2}\ B_1$ *of* H_{-s} $(B_1 = \{f \in L^2(\mu)\ ;\ \|f\|_1 \leq 1\})$ *under condition (*).*

Proof. We see, like Bolthausen ([5]), that this set is $\sqrt{\Gamma}\ (B_s) = \sqrt{2}\ B_1$ using the results of Kuelbs and Lepage ([18]).

3.1. Brownian motion on a compact riemannian manifold.

The space E is here a d-dimensional compact riemannian manifold and $\{X_t\ ;\ t \geq 0\}$ is the brownian motion on E and μ is the uniform measure on E.

LEMMA 7. (i). For $s > d/2$, $\|Z_1\|^2_{-s}$ *is a bounded r.v.*

(ii). $\forall \delta,\ s > 0\ ,\ s > d/2 - 2/(2+\delta) \Rightarrow E\ \|Z_1\|^{2+\delta}_{-s} < \infty$.

Remark : . For $s = d/2$ the r.v. $Z_1 \in H_{-s}$ admits moments of every order.

. For $s > d/2-1$ there is a $\delta > 0$ verifying (ii).

Proof. Note first that $L = -\Delta$, where Δ is the Laplace operator of the Riemann manifold E. The eigenvalues of L, $(\lambda_m)_{m \geq 0}$ satisfy $\lambda_m \sim c\ m^{2/d}$ ([19]) for $m \to \infty$, where $c > 0$ is some constant. From another hand, Giné ([14]) shows :

$$\forall S > d/2 \quad , \quad \exists C > 0 \quad , \quad \forall x \in E \quad , \quad \|\delta_x\|^2_{-S} \leq C$$

(i) $$\|Z_1\|^2_{-s} = \sum_{m=1}^{\infty} \lambda_m^{-s} \left[\int_0^1 e_m(X_u)\, du\right]^2$$

$$\|Z_1\|^2_{-s} \leq 1 + \sum_{m=1}^{\infty} \lambda_m^{-s} \int_0^1 e_m^2(X_u)\, du = \int_0^1 \|\delta_{X_u}\|^2_{-s}\, du \leq C$$

(ii) $$E\,\|Z_1\|^{2+}_{-s} = E\left[\int \sum_{m=1}^{\infty} \lambda_m^{-s} \left(\int_0^1 e_m(X_u)\, du\right)^2\right]^{1+\delta/2}$$

Let $h,k \geq 0$ satisfying $h+k = s$ and $p = (2+\delta)/\delta$, $q = (2+\delta)/2$, Hölder inequality implies $E\,\|Z_1\|^{2+\delta}_{-s} \leq E\,A\,B$ where

$$A = \left[\sum_{m=1}^{\infty} \lambda_m^{-hp} \left(\int_0^1 e_m(X_u)\, du\right)^2\right]^{\delta/2} < C^{\delta/2}$$, because $hp > d/2$, and

$$B = \sum_{m=1}^{\infty} \lambda_m^{-kp} \left(\int_0^1 e_m(X_u)\, du\right)^2$$, and $E\left(\int_0^1 e_m(X_u)\, du\right)^2 \leq 2\,\lambda_m^{-1}$ thus

$$EB \leq 2 \sum_{m=1}^{\infty} \lambda_m^{-(1+kp)} .$$

Setting $hp = kp + 1$, we have $E\,\|Z_1\|^{2+\delta}_{-s} \leq 2 \sum_{m=1}^{\infty} \lambda_m^{-s-2/(2+\delta)}\, C^{\delta/2}$.

Note that Z_n can be rewrite $Z_n = n^{-1/2}\,(Z_1^{(1)}+\ldots+Z_n^{(n)})$, where $Z_1^{(k)} = \sum_{m=1}^{\infty} \left[\int_{k=1}^{k} e_m(X_u)\, du\right] e_m \in H_{-s}$ is a stationary sequence of random variables with $Z_1^{(1)} = Z_1$. Moreover, Baxter and Brosamler ([4]) show that this sequence is ϕ-mixing (with $\phi_n \leq c\,a^n$, $0 < a < 1$). Thus the following theorem 8 will result from lemma 7 and :

THEOREM A. (Dehling, Philipp [8]). Let $\{x_\nu , \nu \geq 1\}$ *a strictly stationary sequence of random variables with values in a separable Hilbert space* H *, centered at expectation and having a* $(2+\delta)$ *-order finite moment* $(0 < \delta \leq 1)$. *If the sequence is strongly mixing with* $\alpha_n = O\,(n^{-(1+\varepsilon)(1+2/\delta)})$ *for* $n\to\infty$ *it can be reconstructed together with a brownian motion* $\{Z(t)\ ;\ t \geq 0\}$ *with covariance* Γ *on another probability space such that :*

$$\Big\|\sum_{\nu\leq t} x_\nu - X(t)\Big\| = o\left(\sqrt{t\,\ell n\,(\ell n\, t)}\right) \quad a.s.$$

Here Γ *is defined by :* $\forall x,y \in H$,

$$(\Gamma x,y) = E(x,x_1)(y,x_1) + \sum_{\nu=2}^{\infty} E\{(x,x_1)(y,x_\nu) + (x,x_\nu)(y,x_1)\} .$$

THEOREM 8. Let $s > d/2-1$ *, the sequence* $Z_n (2\ \ell n(\ell n\ n))^{-1/2}$ *is conditionally compact in* H_{-s} *and its a.s. cluster set for* $n\to\infty$ *is* $\sqrt{2}\ B_1$ *where* B_1 *is the closed unit ball of* H_1 .

There is some $\Omega_0 \in \Omega$ *with* $\mathbb{P}(\Omega_0) = 1$ *and :*

$$\forall\omega \in \Omega_0 , \forall f \in H_s , \overline{\lim_{n\to\infty}} [Z_n(f)(\omega)(2\ \ell n(\ell n\ n))^{-1/2}] = \sqrt{2}\ \|f\|_{-1}$$

Proof : Using lemma 7 we see that theorem A works for the process Z_n ; it allows us to make a strong approximation. With help of proposition we get the first part of theorem. The second part follows from conditional compactness of the sequence and from a.s. convergence of the upper limit for any $f \in H_s$ which is another consequence of theorem applied with $H = \mathbb{R}$ and $x_\nu = Z_1^{(\nu)}(f)$. Those random variable have $(2+\delta)$-order moments because $E|Z_1^{(k)}(f)|^{2+\delta} \leq \|f\|_s^{2+\delta} E\|Z_1^{(k)}\|_{-s}^{2+\delta}$. We construct a sequence $(U_k)_{k\geq 1}$ with same law that $(Z_1^{(k)}(f))_{k\geq 1}$ so that a brownian motion $W(t)$ such that $\sum_{k<t} U_k - W(t)\ a = o(\sqrt{t\ \ell n(\ell n\ t)})$ a.s., $a = \sqrt{2}\ \|f\|_{-1}$. The classical law of iterated logarithm allows us to conclude.

Remarks. . Theorem 8 closes the conjecture (8.11) of [4] for the case of the brownian motion on a compact riemannian manifold.

The case of torus T^d , studied by Bolthausen ([5]) , leads to the same regularity condition $s > d/2-1$. The method used is there martingale theory.

. Conservative diffusions.

If $\{X_t ; t \geq 0\}$ is a diffusion on E with infinitesimal generator $L = -\Delta + V$, for some vector field V . Let $\nu(dx) = \Phi(x)\ \mu(dx)$ the invariant measure of this process (c.f. Ikeda, Watanabe [15]), the function Φ is solution of $L^*\Phi = 0$, where L^* is the adjoint of L in $L^2(\mu)$. If the field V is C^∞ , Φ can be choosen C^∞ and strictly positive. Operator L is self-adjointed in $L^2(\nu)$ if $V = 2\nabla\ \ell n(\Phi)$ (c.f. [15]) . Giné [14] shows that the Sobolev norm with index s associated to L in $L^2(\nu)$ is equivalent to $\|.\|_s$ because norms on $L^2(\nu)$ and $L^2(\mu)$ are equivalent here. If the eigenvalues of L have the same asymptotic behaviour than those of $-\Delta$, then an analogous of theorem 8 is true for such diffusions ; indeed the proof of lemma 7 is still valid here.

3.2. Diffusions on $\mathbb{R}^d$

Results of § 3 are applied to diffusions on $\mathbb{R}^d$ solutions of a stochastic differential equation (S.D.E.) :

$$dX_t = \beta(X_t)\,dt + dW_t$$

Here $\{W_t\ ;\ t \geq 0\}$ is a brownian motion on $\mathbb{R}^d$ and β is a function. Such diffusions have been studied, for example by Albeverio, Hoegh-Krohn and Streit ([2]) and Carmona ([6]). We write C_0^k $(0 \leq k \leq \infty)$ the space of k time differentiable real functions defined on $\mathbb{R}^d$ and with a compact support. Let $\mu(dx) = \Phi^2(x)\,dx$ a absolutely continuous law, we define, like [2] , the bilinear form E on C_0^1 :

$$E\,(f,g) = \int \nabla f \,.\, \nabla g\, d\mu$$

The domain $D(\nabla)$ of E is an Hilbert space with the norm :

$$|f|_1 = (E\,(f,f) + \int f^2\, d\mu)^{1/2} \quad .$$

If the form E is closed, there is a self-adjointed linear operator L on $D(\nabla)$ such that $E(f,g) = (Lf,g)$. It is the case for $\nabla\Phi \in L^2_{loc}(\mathbb{R}^d)$; then $C_0^2 \subset D(L)$, $Lf = -\Delta f - \beta \,.\, \nabla f$ for $f \in C_0^2$ where $\beta = 2\,\Phi^{-1}\,\nabla\Phi$ and $L1 = 0$. If $\Phi^{-1}\,\nabla\Phi,\ \Phi^{-1}\,\Delta\Phi \in L^2_{loc}(\mathbb{R}^d)$ then $\Phi^{-1}\,C_0^\infty \subset D(L)$ and $L(\Phi^{-1}f) = \Phi^{-1}\,Hf$ for $f \in C_0^\infty$ where $H = -\Delta + V$ and $V = \Phi^{-1}\,\Delta\Phi$.

THEOREM B *([2]). Let* $\mu(dx) = \Phi^2(x)\,dx$ *a law equivalent to Lebesgue measure satisfying* $\beta = 2\,\Phi^{-1}\,\nabla\Phi \in L^2(\mu)$ *, there is a unique solution* $\{X_t\ ;\ t \geq 0\}$ *to the S.D.E.:*

$$dX_t = \beta(X_t)\,dt + dW_t \quad .$$

If $\operatorname{div}\beta \in L^2_{loc}(\mathbb{R}^d)$ *satisfies* $\operatorname{div}\beta(x) \geq -c_1\,|x|^2 - c_2$ *for some constants* $c_1, c_2 \geq 0$ *and* $\beta \in L^4_{loc}(\mathbb{R}^d)$, $\Phi^{-1}\,\Delta\Phi \in L^2_{loc}(\mathbb{R}^d)$ *, then the process* $\{X_t\ ;\ t \geq 0\}$ *is markovian with invariant law* μ *.*

The operators H and L with domains $D(H) \subset L^2(\mathbb{R}^d)$, $D(L) \subset L^2(\mu)$ have the same spectrum. It is discrete if $\lim_{|x|\to\infty} V(x) = +\infty$.

Condition (*) is satisfied with the help of Tamura's result ([24]) which asserts that $\lim_{k\to\infty} \lambda_k^{d/2+d/m}\, k^{-1} = c > 0$ under assumption A_m :

$$A_m \begin{cases} \text{i.} & \exists R > 0 \ , \ 0 < \text{Inf}\,\{V(x)\ ;\ |x| > R\} \leq \text{Sup}\,\{V(x)\,|x|^{-m}\ ;\ |x| > R\} < \infty \\ \text{ii.} & \forall \alpha = (\alpha_1,\dots,\alpha_d) \in \mathbb{N}^d \ , \ \exists C_\alpha > 0 \ , \ \forall x \in \mathbb{R}^d \ , \ a = \alpha_1 + \dots + \alpha_d \\ & |D^\alpha V(x)| \leq C_\alpha\,(1 + |x|^2)^{(m-a)/2} \\ \text{iii.} & \exists C > 0 \ , \ |x| > R \Rightarrow x.V(x) \geq C\,|x|^m \end{cases}$$

COROLLARY 9. If μ *is equivalent to Lebesgue measure,* $V = \Phi^{-1}\,\Delta\Phi$ *satisfies assumptions* A_m *for some* $m > 0$ *and* $\beta = 2\,\Phi^{-1}\,\nabla\Phi \in L^2(\mu) \cap L^4_{loc}(\mathbb{R}^d)$, *then* $s > d/2 + d/m - 1$ *implies that the sequence* Z_n *converges in distribution to* Z *in* H_{-s} .

Remark : Conditions A_m are satisfied by homogeneous polynomials V with degree m .

A multidimensional Ornstein-Uhlenbeck (i.e. $V(x) = c\,|x|^2$ for some constant $c > 0$) satisfies hypotheses of Corollary 9 if $s > d-1$.

We do not get here an iterated logarithm law (L.I.L.) because the lemma of the former section is no longer valid. To avoid this problem and obtain a uniform L.I.L. we now reduce the class of functions used. From here we suppose Φ bounded, thus $L^2(\mathbb{R}^d) \subset L^2(\mu)$; the Sobolev space of $L^2(\mathbb{R}^d)$ constructed with tensor products of Hermite functions is denoted by H^*_s and its norm by $\|\cdot\|^*_s$. The process Z_1 is a bounded random variable with values in H^*_{-s} for $s > d-1/6$:

$$\|Z_1\|^{*2}_{-s} = \sum_{\nu\ \mathbb{N}^d} \theta_\nu^{-s} \left(\int_0^1 H_\nu\,(X_u)\,du\right)^2$$

Here $H_\nu\,(x_1,\dots,x_d) = h_{\nu_1}(x_1)\dots h_{\nu_d}(x_d)$, $\theta_\nu = \sum_{j=1}^{d} (\nu_j+1)$ for $\nu = (\nu_1,\dots,\nu_d)$,

and h_m is the m-th normalized Hermite function. With the help of the estimate $\|h_m\|_\infty = O\,(m^{-1/12})$ of Szégö ([23]) we see, summing by parts on spheres of $\mathbb{N}^d$, that : $\|Z_1\|^{*2}_{-s} \leq C\,d^{1-s} \sum_{p=1}^{\infty} p^{d-s-7/6}$.

The strongly mixing property of the $\{X_t\ ;\ t \geq 0\}$ process implies, by an analogous method than for theorem 8,

THEOREM 10. For $s > d - 1/6$ *, there is some* $\Omega_o \subset \Omega$ *with* $\mathbb{P}(\Omega_o) = 1$ *such that :*

$$\forall \omega \in \Omega_0 \ , \ \forall f \in H_s^* \ , \ \overline{\lim_{n\to\infty}} \left[(Z_n(f)\,(\omega) - \sqrt{n}\int f\, d\mu)\,(2\ \ell n(\ell n\ n))^{-1/2}\right] = \sqrt{2}\,\|f\|_{-1}$$

The use of the norm $\|\cdot\|_{-1}$ *is valid here because of inclusion* $H_s^* \subset H_{-1}$.

Proof : Note that $H_s^* \subset L^\infty(\mathbb{R}^d)$ thus $f \in H_s^*$ satisfies integrability condition $\int |f|^{2+\delta} d\mu < \infty$ of ([3], theorem 2.7) and individual L.I.L. is satisfied.

From the other hand, theorem A still applies. The process Z_n satisfies a strong invariance principle with speed $o\,((\ell n\,(\ell n\ n))^{-1/2})$ in H_{-s}^* . The limit process Z satisfies $E(Z,H_\nu)_{-s}^{*2} = \theta_\nu^{-2s}\,(G(H_\nu - \bar{H}_\nu)\ ,\ H_\nu - \bar{H}_\nu)$ where $\bar{H}_\nu = \int H_\nu(x)\,\mu(dx)$ for $\nu \in \mathbb{N}^d$; its covariance is a trace class operator because $\tilde{H}_\nu = \theta_\nu^{s/2}\,H_\nu$ is a orthonormal basis of H_s^* and :

$$\sum_{\nu\in\mathbb{N}^d} E(Z,\tilde{H}_\nu)_{-s}^{*2} = \sum_{\nu\in\mathbb{N}^d} \theta_\nu^{-s}\,(G(H_\nu - \bar{H}_\nu)\ ,\ H_\nu - \bar{H}_\nu) \leq 2\,\lambda_1^{-1}\int \|\delta_x\|_{-s}^{*2}\,\mu(dx) < \infty\ .$$

Thus we get a compact subset K of H_{-s}^* associated to the gaussian law of Z. This compact set is the a.s. cluster set of $(Z_n - EZ_n)/\sqrt{2\ \ell n(\ell n\ n)}$.

The theorem follows .

Remark. Note that $C_0^\infty \subset H_s^*$, so that this theorem establishes a uniform L.I.L. on the class C_0^∞. This result is connected with the conjecture (8.11) of [4] .

3.3. Discretization.

Suppose here that δ_{X_t} is a bounded random variable of H_{-s} ; for example if $s > d/2$ and if $\{X_t\ ;\ t \geq 0\}$ is a diffusion on a compact riemannian d-dimensional manifold this condition is realized.

The discretization of the process Z_n is, for $\Delta > 0$:

$$Z_{n,\Delta}\,(f) = \frac{1}{\sqrt{n}} \sum_{1\leq k\Delta\leq n} f\,(X_{k\Delta})\ , \qquad f \in H_s$$

For $\Delta > 0$ fixed, we see that [8] implies $Z_{n,\Delta}$ converge in distribution to a gaussian process Z_Δ such that :

$$\forall f \in H_s \quad E(Z_\Delta(f))^2 = \Delta \sum_{m=1}^{\infty} k^2\,(\lambda_m\Delta)\,(f,e_m)^2\ , \quad \text{if}\ \ k(v) = \left(\frac{1+e^{-v}}{1-e^{-v}}\right)^{1/2}$$

With the help of an i.i.d. sequence of normal realizations $(\xi_m)_{m\geq 1}$ we give explicit constructions of Z and Z_Δ :

$$Z = \sum_{m=1}^{\infty} \sqrt{\frac{2}{\lambda_m}}\, \xi_m\, e_m \quad , \qquad Z_\Delta = \sqrt{\Delta} \sum_{m=1}^{\infty} k(\lambda_m \Delta)\, \xi_m\, e_m$$

Thus $E\,\|Z_\Delta - Z\|^2_{-s} = \Delta \sum_{m=1}^{\infty} \left|k(\lambda_m\Delta) - \sqrt{\frac{2}{\lambda_m \Delta}}\right|^2 \lambda_m^{-s} \leq \Delta \sum_{m=1}^{\infty} \{2 \wedge (\frac{\lambda_n \Delta}{3})\}\, \lambda_m^{-s}$

For a diffusion on a compact riemannian manifold :

$$E\,\|Z_\Delta - Z\|^2_{-s} \leq C\, \Delta^{s-d/2+1} \quad \text{for } \Delta \to 0, \text{ if } s > d/2 \ .$$

Then the Prohorov and Levy's distances of those gaussian random variables on H_{-s} are $O(\Delta^{D/3})$ for $\Delta \to 0$, $D = s - d/2+1$.

From the other hand, a precise analysis of the results of [11] shows that Dudley distance d_3 can be estimate :

$$d_3\,(\mathbb{P}_{Z_{n,\Delta}}, \mathbb{P}_{Z_\Delta}) = O\,(\Delta^{-11/4}\, n^{-1/4}) \text{ for } n \to \infty \ , \ \Delta \to 0 \quad \text{if} \quad s > d/2.$$

Thus the discretized process $Z_{n,\Delta(n)}$ converge in distribution to Z for $\Delta\,(n) = o\,(n^{-1/11})$ when $n \to \infty$. If, moreover $\Delta(n) = O\,(n^{-1/(4D+11)})$ then :

$$d_3\,(\mathbb{P}_{Z_{n,\Delta(n)}}, \mathbb{P}_Z) = O\,(n^{-D/(4D+11)}) \quad .$$

For great values of s this speed is approximately $n^{-1/4}$.

BIBLIOGRAPHY :

[1] A. De Acosta. *Existence and convergence of probability measures on Banach spaces. Trans. Amer. Math. Soc. 152, pp. 273-298 (1970).*

[2] S. Albeverio, R., Hoegh-Krohn, L. Streit. *Energy forms, Hamiltonian and distorted Brownian paths. J. of Math. Phys. 18, n° 5, pp. 907-917 (1977).*

[3] R.M. Battacharya. *On the functionnal central limit theorem and the law of the iterated logarithm for Markov processes. Zeit. für Wahr. und Verw. Gebiete 60, pp. 185-201 (1982).*

[4] J.R. Baxter, G.A. Brosamler. *Energy and the law of the iterated logarithm. Math. Scand. 38, pp. 115-136 (1976).*

[5] E. Bolthausen. *On the asymptotic behaviour of the empirical random field of the Brownian motion. Stoch. Pr. and their Appl. 16, pp. 199-204,(1983).*

[6] R. Carmona. *Processus de diffusion gouverné par la forme de Dirichlet de l'opérateur de Schrödinger. Séminaire de probabilité XIII, Strasbourg 1977-1978, L.N.M. 721, pp. 557-569 (1979).*

[7] D. Dacunha-Castelle, D. Florens. *Choix du paramètre de discrétisation pour estimer le paramètre d'une diffusion. C.R.A.S. Série I, Paris, t.299, pp. 65-69 (1984).*

[8] H. Dehling, W. Philipp. *Almost sure invariance principles for weakly dependent vector-valued random variables. Ann. of Prob. 10, pp. 689-701 (1982).*

[9] P. Doukhan. *Fonctions d'Hermite et statistiques des processus mélangeants (Submitted to publication, 1985).*

[10] P. Doukhan, J.R. Leon, F. Portal . *Principe d'invariance faible pour la mesure empirique d'une suite de variables aléatoires dépendantes. (Submitted to publication, 1985).*

[11] P. Doukhan, J.R. Leon, F. Portal. *Calcul de la vitesse de convergence dans le théorème central limite vis à vis des distances de Prohorov, Dudley et Levy dans le cas de variables aléatoires dépendantes. Prob. and Math. Stat. VI.2 , 1985.*

[12] X. Fernique. *Régularité des trajectoires des fonctions aléatoires gaussiennes.* L.N.M. 480, Springer (1975).

[13] D. Florens. *Théorème de limite centrale des fonctionnelles de diffusions.* C.R.A.S. Série I, Paris, t. 299, pp. 995-998 (1984).

[14] E. Giné. *Invariant test for uniformity on compact riemannian manifolds based on Sobolev norms.* Ann. of Stat. 3, pp. 1243-1266 (1975).

[15] N. Ikeda, S. Watanabe. *Stochastic differential equations and diffusion processes.* North-Holland, Tokyo (1981).

[16] K. Ito, H.P. Mc Kean. *Diffusion processes and their sample paths.* Springer Verlag, Berlin (1974).

[17] J. Kuelbs. *Kolmogorov law of the iterated logarithm for Banach space valued random variables.* Illinois J. of Math. 21, pp. 784-800 (1977).

[18] J. Kuelbs, R. Lepage. *The law of the iterated logarithm for Brownian motion in a Banach space.* Trans. of the Amer. Math. Soc. 185, pp.253-264 (1973).

[19] S. Minakshisundaram, A. Pleijel. *Some properties of the eigenfunctions of the Laplace-operator on riemannian manifolds.* Can. J. of Math. 1, pp. 242-256 (1943).

[20] L. Nirenberg. *Pseudodifferential operators.* Proc. of Symp. in pure Math. XVI. Global Analysis, A.M.S. Providence, pp. 149-167 (1970).

[21] M. Rosenblatt. *Markov processes.* Springer Verlag, New York (1971).

[22] R.T. Seeley. *Complex powers of an elliptic operator.* Proc. Symp. in pure Math. X, A.M.S., Providence, pp. 288-307 (1968).

[23] G. Szegö. *Orthogonal polynomials.* A.M.S. Providence (1939).

[24] H. Tamura. *Asymptotic formulas with sharp remainder estimates for eigenvalues of elliptic operators of second order.* Duke Math. J. 49, pp. 87-119 (1982).

ALMOST EXCHANGEABLE SEQUENCES IN L^q , $1 \le q < 2$.

Sylvie GUERRE
Equipe d'Analyse
U.A. N° 754
Université Paris VI
4 place Jussieu
75230 PARIS CEDEX 19
Tour 46/0 - 4ème étage

Let $(\Omega,\mathcal{B},P)$ be a probability space and $L^q = L^q(\Omega,\mathcal{B},P)$, $1 \le q < +\infty$. We consider the two following problems :

Problem 1 : Let $(X_n)_{n\in\mathbb{N}}$ be a weakly null sequence in L^q , $1 \le q < +\infty$. Does there exist a subsequence $(X_{n_k})_{k\in\mathbb{N}}$ of $(X_n)_{n\in\mathbb{N}}$ which is almost symmetric ?

[$(X_{n_k})_{k\in\mathbb{N}}$ is almost symmetric if :

$\forall \varepsilon > 0$, $\exists k \in \mathbb{N}$ such that : $\forall (\alpha_i) \in \mathbb{R}^{(\mathbb{N})}$, $\forall \pi$ permutation of $\mathbb{N}$,

$$(1-\varepsilon) \|\Sigma \alpha_i X_{n_{k+i}}\| \le \|\Sigma \alpha_i X_{n_{k+\pi(i)}}\| \le (1+\varepsilon) \|\Sigma \alpha_i X_{n_{k+i}}\|] .$$

Problem 1 was stated for $q > 2$ and solved for $q \in 2\mathbb{N}$ in [8] .

Problem 2 : Under the same hypothesis, does there exist a positive density φ , a subsequence $(X_{n_k})_{k\in\mathbb{N}}$ of $(X_n)_{n\in\mathbb{N}}$ and a exchangeable sequence $(Z_k)_{k\in\mathbb{N}}$ in $L^q(\varphi\, dP)$ such that :

$$\sum_{k=0}^{+\infty} \left\| Z_k - \frac{X_{n_k}}{\varphi^{1/q}} \right\|_{L^q(\varphi dP)} < +\infty \ ?$$

[We will say that $(X_n)_{n\in\mathbb{N}}$ is almost exchangeable after the change of density φ . If in addition $(Z_k)_{k\in\mathbb{N}}$ is i.i.d., we will say that $(X_n)_{n\in\mathbb{N}}$ is almost i.i.d. after the change of density φ] .

This definition comes from [2] .

The following implications are well known :

$(X_{n_k})_{k\in\mathbb{N}}$ is almost i.i.d. after the change of density φ

$\Downarrow$

$(X_{n_k})_{k\in\mathbb{N}}$ is almost exchangeable after the change of density φ

$\Downarrow$

$(X_{n_k})_{k\in\mathbb{N}}$ is almost symmetric.

By the Finetti's theorem (cf. [2]), we also have that :

$(X_{n_k})_{k\in\mathbb{N}}$ is almost exchangeable

$\Downarrow$

$(X_{n_k})_{k\in\mathbb{N}}$ is almost i.i.d., conditionnally to its tail field.

On the contrary, a modification of Example 2 in [2] by Y. Raynaud shows that there exists an almost symmetric sequence in L^q with no almost exchangeable subsequence after any change of density.

A lot of results are known about symmetric subspaces of L^q , $1 \le q < +\infty$, in particular about those that are isomorphic to ℓ^p for some p . In almost all these papers, symmetric sequences are in fact i.i.d. or at least exchangeable : this is the most natural way to find symmetric sequences in L^q-spaces. I recall here the results that are closely related to the problems 1 and 2 and that motivated these questions.

Kadec-Pelczynski [9].

For $q > 2$, every weakly null sequence in L^q :

- either is isomorphic to the unit vector basis of ℓ^2
- or has a subsequence which is almost equivalent to the unit vector basis of ℓ^q .

Dacunha-Castelle [4].

For $1 \le q < 2$, 1-symmetric subspaces of L^q are means of Orlicz spaces.

D. Aldous [1]. (Case $q = 1$).

J.L. Krivine and B. Maurey [10].

Every infinite dimensional subspace of L^q , $1 \le q < +\infty$, has a subspace which is isomorphic to ℓ^p for some p .

More precisely : If $(X_n)_{n\in\mathbb{N}}$ is a weakly null sequence in L^q , $1 \le q < +\infty$, there exists $p \in [1,q]$ such that for all $\varepsilon > 0$ there exists a sequence of blocks $(Y_n)_{n\in\mathbb{N}}$ on $(X_n)_{n\in\mathbb{N}}$ such that $(Y_n)_{n\in\mathbb{N}}$ is $(1+\varepsilon)$-equivalent to the unit vector basis of ℓ^p and almost i.i.d. after a change of density.

This last result answers positively to problem 1 and 2 if we are allowed to take blocks on $(X_n)_{n\in\mathbb{N}}$ and not only subsequences.

Recall now partial known positive answer to problem 1 :

THEOREM 1. If $q > 2$, every weakly null sequence of L^q has an almost symmetric subsequence.

If $1 \le q < 2$, every sequence in L^q , which is equivalent to the unit vector basis of ℓ^2 has an almost symmetric subsequence.

The case $q = 1$ is due to H.P. Rosenthal.

The case $q \in 2\mathbb{N}$ is proved in [8].

The proof of the general case [6] uses the theory of stability [10] : in stable spaces, there is a natural way to find almost symmetric sequences. First, recall a few definitions :

A Banach space X is <u>stable</u> if given two bounded sequences $(x_n)_{n\in\mathbb{N}}$ and $(y_m)_{m\in\mathbb{N}}$ in X and two ultrafilters $\mathcal{U}$ and $\mathcal{V}$ on $\mathbb{N}$, we have :

$$\lim_{n,\mathcal{U}} \lim_{m,\mathcal{V}} \|x_n + y_m\| = \lim_{m,\mathcal{V}} \lim_{n,\mathcal{U}} \|x_n + y_m\| .$$

The <u>type</u> σ defined by $(x_n)_{n\in\mathbb{N}}$ and $\mathcal{U}$ is a function from X to $\mathbb{R}^+$ such that :

$$\forall\, x \in X,\quad \sigma(x) = \lim_{n,\mathcal{U}} \|x + x_n\| .$$

For $(\alpha,\beta) \in \mathbb{R}^2$, we define the <u>type</u> $\alpha\sigma * \beta\tau$ by :

$$\forall\, x \in X \qquad \alpha\sigma * \beta\tau(x) = \lim_{n,\mathcal{U}} \lim_{m,\mathcal{V}} \|x + \alpha x_n + \beta y_m\|$$

where σ is defined by $(x_n)_{n\in\mathbb{N}}$ and $\mathcal{U}$ and τ by $(y_m)_{m\in\mathbb{N}}$ and $\mathcal{V}$.
Let $\|\sigma\| = \sigma(0) = \lim_{n,\mathcal{U}} \|x_n\|$.

The <u>spreading model</u> [3] defined by $(x_n)_{n\in\mathbb{N}}$ and $\mathcal{U}$ is the completion of $\mathbb{R}^{(\mathbb{N})}$ under the norm :

$$\Big\| \sum_{i=1}^{k} \alpha_i \ell_i \Big\| = \lim_{n_1} \ldots \lim_{n_k} \| \alpha_1 x_{n_1} + \ldots + \alpha_k x_{n_k} \| .$$

($(x_n)_{n\in\mathbb{N}}$ is supposed to have no convergent subsequences).

In a stable space, every spreading model is 1-symmetric. The proof of theorem 1 uses a sufficient condition (S.C.) in stable spaces for a sequence $(x_n)_{n\in\mathbb{N}}$ to have a subsequence which is almost equivalent to the fundamental sequence of its spreading model :

(S.C.) : Let σ be the type defined by $(x_n)_{n\in\mathbb{N}}$ on a stable space X. If $K_1(\sigma) = \{\tau / \exists\, k \in \mathbb{N},\ \exists\, \alpha_1,\ldots,\alpha_k \in \mathbb{R}^k$ such that $\tau = \alpha_1\sigma * \ldots * \alpha_k\sigma$ and $\|\tau\| \le 1\}$ is relatively compact for the uniform convergence on bounded sets of X, then $(x_n)_{n\in\mathbb{N}}$ has a subsequence which is almost equivalent to the spreading model defined by $(x_n)_{n\in\mathbb{N}}$ and thus almost symmetric.

This condition was used by J.L. Krivine and B. Maurey in the case of ℓ^p-types and is written in [6]. Its proof uses Ascoli's theorem.

Let now $(X_n)_{n\in\mathbb{N}}$ be a weakly null sequence in L^q $1 \le q < +\infty$ which is equivalent to the unit vector basis of ℓ^2 (even in the case $q > 2$, this is the only case to consider because of Kadec-Pelczynski's result [9]). It is shown in [6] that $(X_n)_{n\in\mathbb{N}}$ verifies (S.C.) : suppose for simplicity that $1 \le q < 2$ and that the type σ defined by $(X_n)_{n\in\mathbb{N}}$ is symmetric [i.e. : $\forall\, X \in L^q$, $\sigma(X) = \sigma(-X)$]. One can show that σ is entirely determined by $\|\sigma\|$ and a function U^σ belonging to $L^2(\Omega \times [0,+\infty[\,, dP \otimes \frac{dt}{t^{q+1}})$ which is the weak limit in that space of $(1 - \cos t\, X_n)_{n\in\mathbb{N}}$, by the formula : $\forall\, X \in L^q$,

$$K_q(\sigma(X)^q - \|\sigma\|^q - \|X\|^q) = -\langle U^\sigma, U^X \rangle \quad \text{where} \quad K_q = \int_0^{+\infty} (1 - \cos t) \frac{dt}{t^{q+1}} ,$$

$U^X = 1 - \cos t\, X$ and $<,>$ is the inner product in $L^2(dP \otimes \frac{dt}{t^{q+1}})$.

Moreover, this representation has the following property :

$$\left\{ \begin{array}{l} \sigma_n \xrightarrow[n\to+\infty]{} \sigma \text{ uniformly on bounded sets of } L^q \\ \Leftrightarrow \left\{ \begin{array}{l} u^{\sigma_n} \xrightarrow[n\to+\infty]{} u^{\sigma} \quad \text{in } L^2(dP \otimes \frac{dt}{t^{q+1}}) \\ \|\sigma_n\| \xrightarrow[n\to+\infty]{} \|\sigma\| \ . \end{array} \right. \end{array} \right.$$

To prove theorem 1, it is thus sufficient to show that if $\tau_n = \alpha_1^n \sigma * \dots * \alpha_{k_n}^n \sigma$ belongs to $K_1(\sigma)$ then $(U^{\tau_n})_{n\in\mathbb{N}}$ has a subsequence which converges in $L^q(dP \otimes \frac{dt}{t^{q+1}})$. As $(X_n)_{n\in\mathbb{N}}$ is equivalent to ℓ^2 , one can show that : $\exists\, A \in L^1$ such that :

$$U^{\sigma}(\omega,t) = 1 - e^{-t^2 A(\omega)} + t^2\, \varepsilon(\omega,t)$$

where $\lim_{t\to 0} \varepsilon(\omega,t) = 0$ a.e.

On the other hand, we know that :

$$1 - U^{\tau_n} = \prod_{i=1}^{k_n} (1 - U^{\alpha_i^n \sigma}) \qquad [6] \ .$$

We deduce from these two facts that :

$$U^{\tau_n} = 1 - e^{-\sum_{i=1}^{k_n} (\alpha_i^n t)^2 A(\omega)} + \sum_{i=1}^{k_n} (\alpha_i^n)^2\, \varepsilon(\omega, S_n t)$$

where $S_n = \operatorname{Sup}\{|\alpha_i^n|,\ 1 \le i \le k_n\}$.

Taking a subsequence of $(X_n)_{n\in\mathbb{N}}$, we can suppose :

$$\left\{ \begin{array}{l} \sum_{i=1}^{k_n} (\alpha_i^n)^2 \xrightarrow[n\to+\infty]{} \alpha > 0 \\ S_n \xrightarrow[n\to+\infty]{} 0 \ . \end{array} \right.$$

Then $(U^{\tau_n})_{n\in\mathbb{N}}$ converges a.e. to $1 - e^{-\alpha t^2 A(\omega)}$ and by Lebesgue's dominated convergence theorem $(U^{\tau_n})_{n\in\mathbb{N}}$ converges also in $L^2(dP \otimes \frac{dt}{t^{q+1}})$. This implies that $K_1(\sigma)$ is relatively compact for the uniform convergence on bounded sets of L^q and proves theorem 1 by (S.C.) in that case.

This theorem does not give any answer to problems 1 or 2 for weakly null sequences of L^q $1 \le q < 2$ which are not equivalent to the unit vector basis of ℓ^2 . The following result gives a negative answer to problem 2 in that case :

THEOREM 2. <u>Let</u> $1 \le q < p < 2$. <u>There exists a bounded sequence in</u> L^q <u>such that</u> :

(i) $(X_n)_{n\in\mathbb{N}}$ <u>is equivalent to the unit vector basis of</u> ℓ^p .

(ii) $(X_n)_{n\in\mathbb{N}}$ has no almost exchangeable subsequence after any change of density.

(iii) If σ is the type defined by $(X_n)_{n\in\mathbb{N}}$ $K_1(\sigma)$ is not relatively compact for the uniform convergence on bounded sets of L^q.

Remark : Property (iii) gives a hope that $(X_n)_{n\in\mathbb{N}}$ has no almost symmetric subsequence but this question is still open. In fact, the two natural ways to find almost symmetric subsequences in L^p-spaces (namely the theory of probability with almost exchangeable subsequences and the stability of those spaces) do not work for this sequence $(X_n)_{n\in\mathbb{N}}$.

Sketch of proof [7].

$$\text{Let } U(\omega,t) = 1 - e^{-\int_0^{+\infty}(1-\cos t\,u)\,\chi(u,\omega)\,\frac{du}{u^{p+1}}}$$

$$\text{where } \chi(u,\omega) = 1 \quad \text{if } u\in[0,1/N]$$
$$= \xi_k(\omega) \quad \text{if } u\in[N^{2k-1},N^{2k+1}]$$

[N is a fixed positive constant and $(\xi_k)_{k\in\mathbb{N}}$ a sequence of i.i.d. random variables such that $P(\xi_k=1)=P(\xi_k=2)=1/2$ and $\sigma(\xi_k, k\in\mathbb{N}) = \mathcal{B}$] .

It is possible to verify that U defines a symmetric type σ on L^q (cf. [11]) by the way defined in theorem 1 and such that :

$$\begin{cases} U^\sigma = U \\ K_q\|\sigma\|^q = \|U^\sigma\|_{L^1(dP\otimes \frac{dt}{t^{q+1}})} . \end{cases}$$

By construction, the function U oscilates a lot [$(\xi_k)_{k\in\mathbb{N}}$ is i.i.d. and so has no a.e. convergent subsequence] and this prevents $K_1(\sigma)$ to be uniformly relatively compact and proves property (iii).

On the other hand, it is easy to see that U satisfies the inequalities :

$$1-e^{-K_p t^p} \le U^\sigma(\omega,t) \le 1-e^{-2K_p t^p}$$

$$\left(\text{where } K_p = \int_0^{+\infty}(1-\cos u)\frac{du}{u^{p+1}}\right) .$$

Using techniques of stability, it is possible to show that these inequalities imply that σ is "equivalent" to an ℓ^p-type τ (such that $U^\tau = 1-e^{-K_p t^p}$) and thus can be defined by a sequence $(X_n)_{n\in\mathbb{N}}$ which is equivalent to the unit vector basis of ℓ^p. This proves property (i).

Proof of (ii) : We follow an idea of [2].

Suppose that there exists a density φ, an exchangeable sequence $(Z_k)_{k\in\mathbb{N}}$ in $L^q(\varphi dP)$ and a subsequence $(Xn_k)_{k\in\mathbb{N}}$ of $(X_n)_{n\in\mathbb{N}}$ such that :

$$\sum_{k=1}^{+\infty} \| Z_k - \frac{X_{n_k}}{\varphi} \|_{L^q(\varphi dP)} < +\infty \quad .$$

Then, it is shown in [2] that :

$$U(\omega, \frac{1}{\varphi(\omega)^{1/q}}) = w - \lim_{k\to+\infty} (1 - e^{it \frac{X_{n_k}}{\varphi^{1/q}}})$$

$$= w - \lim_{k\to+\infty} (1 - e^{it Z_k})$$

$$= E^{\mathcal{G}} [1 - e^{it Z_k}] \quad \text{for all } k \in \mathbb{N} \ .$$

Where $\mathcal{G}$ is the σ-field generated in Ω by the functions $\omega \to U(\omega, \frac{t}{\varphi(\omega)^{1/q}})$ for $t \in \mathbb{R}^+$ ($\mathcal{G}$ is smaller than the tailfield of $(Z_k)_{k\in\mathbb{N}}$) .

<u>Let us show that</u> $\mathcal{G} = \mathcal{B}$: $\mathcal{G}$ is also generated by :

$$\omega \to W_t(\omega) = \int_0^{+\infty} [1 - \cos(\frac{t}{\varphi(\omega)^{1/q}} u)] \ \chi(u,\omega) \ \frac{du}{u^{p+1}} \quad .$$

$$= \int_0^{1/N} [1 - \cos(\frac{t}{\varphi(\omega)^{1/q}} u)] \frac{du}{u^{p+1}} + \sum_{k=0}^{+\infty} \xi_k(\omega) \int_{N^{2k-1}}^{N^{2k+1}} (1 - \cos \frac{1}{\varphi(\omega)^{1/q}} u) \frac{du}{u^{p+1}} \quad .$$

When $t \to +\infty$, it is easy to see that :

$$\begin{cases} \displaystyle\int_0^{1/N} (1 - \cos \frac{t}{\varphi(\omega)^{1/q}} u) \frac{du}{u^{p+1}} \overset{c}{\sim} \frac{t^p}{\varphi(\omega)^{p/q}} \\ \displaystyle\int_{N^{2k-1}}^{N^{2k+1}} (1 - \cos \frac{t}{\varphi(\omega)^{p/q}} u) \frac{du}{u^{p+1}} \to \int_{N^{2k-1}}^{N^{2k+1}} \frac{du}{u^{p+1}} < +\infty \end{cases}$$

for $k \in \mathbb{N}$.

This proves that φ belongs to $\mathcal{G}$ and ξ_k also for all k because we can write :

$$\int_{N^{2k+1}}^{N^{2k+3}} \frac{du}{u^{p+1}} < 2 \int_{N^{2k+1}}^{N^{2k+3}} \frac{du}{u^{p+1}} < \int_{N^{2k-1}}^{N^{2k+1}} \frac{du}{u^{p+1}} \quad .$$

As $(\xi_k)_{k\in\mathbb{N}}$ generates $\mathcal{B}$, $\mathcal{G}$ is equal to $\mathcal{B}$.

This situation is impossible because that would mean that :

$$U(\omega, \frac{t}{\varphi(\omega)^{p/q}}) = 1 - e^{it Z_k} \quad \text{for all } k \in \mathbb{N} \ ,$$

and this is obviously false (for example, $U(\omega, \frac{t}{\varphi(\omega)^{p/q}}) \underset{t\to 0}{\sim} \frac{t^p}{\varphi(\omega)^{p/q}})$.

This proves property (ii) and theorem 2.

BIBLIOGRAPHIE

[1] ALDOUS D.J. : Subspaces of L^1 via random measures.
Trans. Amer. Math. Soc., 267, (1981), 445-463.

[2] BERKES I. and ROSENTHAL H.P. : Almost exchangeable sequences of random variables.
To appear in Zeitschrift für Wahrscheinlichkeitstheorie verw. Gebiete.

[3] BRUNEL A. and SUCHESTON L. : On B-convex Banach spaces.
Math. Systems theory, t. 7 n°4, 1973.

[4] DACUNHA-CASTELLE D. : Variables aléatoires échangeables et espaces d'Orlicz.
Séminaire Maurey-Schwartz, Ecole Polytechnique, 1974/75, exposés 10 et 11 .

[5] DACUNHA-CASTELLE D. et KRIVINE J.L. : Sous-espaces de L^1 .
Israël Journal of Math., 26 (1977), 320-351.

[6] GUERRE S. : Types et suites symétriques dans L^p , $1 \leq p < +\infty$.
A paraître dans Israël Journal of Math.

[7] GUERRE S. : Sur les suites presque échangeables dans L^q , $1 \leq q < 2$.
Preprint.

[8] JOHNSON W.B., MAUREY B., SCHECHTMAN G., TZAFRIRI L. : Symmetric structures in Banach spaces.
Memoirs of the American Math. Soc., May 1979, vol. 19, n° 217.

[9] KADEC H.I. and PELCZYNSKI A. : Bases, lacunary sequences and complemented subspaces of L_p .
Studia Math., T XXI, 1962.

[10] KRIVINE J.L. et MAUREY B. : Espaces de Banach stables.
Israël Journal of Math., vol. 39, n° 4, (1981).

[11] LEVY P. : Théorie de l'addition des variables aléatoires.
Gauthier-Villars.

AN APPLICATION OF A MARTINGALE INEQUALITY OF DUBINS AND FREEDMAN TO THE LAW OF LARGE NUMBERS IN BANACH SPACES

Bernard HEINKEL
Département de Mathématique
7, rue René Descartes
67084 STRASBOURG Cédex (France)

ABSTRACT : In a real, separable, p-uniformly smooth Banach space the law of large numbers in the Prohorov setting is studied by a method depending on a result of Dubins and Freedman which compares the distribution of a real valued martingale with the one of the associated conditional variances. Some laws of large numbers of Kolmogorov-Brunk type are also given.

Several recent papers have improved very much the knowledge on the strong law of large numbers (SLLN) and on the law of the iterated logarithm (LIL) for random variables (r. v.) taking their values in a Banach space equipped with a regular norm. The key idea in these papers is to use the regularity of the norm for reducing the infinite dimensional SLLN or LIL to a scalar SLLN or LIL. But then for solving this finite dimensional SLLN or LIL problem, sophisticated techniques - multiple truncations, iteration of martingale exponential inequalities - have to be used.

Here we will introduce a new approach of the SLLN in a Banach space with a regular norm, approach which allows both to obtain new results in the non i. i. d. setting and also to show in a simpler way statements which were known previously. The cornerstone of this method is a result of Dubins and Freedman [3] which compares the distribution of a real valued martingale and the ones of the associated conditional variances. Before to state and to prove the results we recall some definitions.

§ 1. SOME DEFINITIONS.

In all the sequel we will denote by $(B, \| \ \|)$ a real separable Banach space which is p-uniformly smooth $(1 < p \leq 2)$; this means that its modulus of smoothness ρ :

$$\forall t > 0, \quad \rho(t) = \sup\left(1/2 \left(\|x+ty\| + \|x-ty\| \right) - 1 \ , \ \|x\| = \|y\| = 1 \right),$$

satisfies :

$\rho(t) \leq C\, t^p$.

C being a positive constant.

It is well known that the norm $\|\ \|$ is differentiable away from the origin ; let's denote by D the derivative of $\|\ \|$.

If now one associates to D the following fonction $F : B \to B'$:

$\forall x \neq 0$, $F(x) = \|x\|^{p-1} D(x/\|x\|)$,

and : $F(0) = 0$,

one can check that F has the following two properties [19] :

(i) $\|F(x)\|_{B'} = \|x\|_B^{p-1}$.

(ii) $\exists\, C_p > 0 : \forall (x,y) \in B^2$, $\left| \|x+y\|^p - \|x\|^p \right| \leq p\, |F(x)(y)| + C_p \|y\|^p$.

These two properties are crucial for reducing the infinite dimensional SLLN to a scalar one. Other geometrical properties of a p-uniformly smooth space that we will use are its reflexivity [5] and the fact that it is of type p [15] .

Usually the SLLN problem in $(B, \|\ \|)$ is stated in the following way :

"Let (X_k) be a sequence of independent, centered, B-valued r.v. and denote by $S_n = X_1 + \ldots + X_n$ the associated partial sums ; under what hypotheses does (S_n/n) converge a.s. to 0 ? (1) "

This point of view is very restrictive because two other asymptotic behaviours of the sequence (S_n/n) are worth of interest :

$P(\sup_n \|S_n/n\| < +\infty) = 1$, which behaviour can be called a bounded law of large numbers (BLLN) (2) ,

and :

$P(\omega : S_n(\omega)/n \to 0 \text{ weakly}) = 1$, which is a law of large numbers in the weak topology (WTLLN) (3) .

The main goal of this paper is to study the three forms (1), (2) and (3) of the LLN, under Prohorov boundedness conditions of the r.v., in p-uniformly smooth spaces ; the good geometrical properties of these spaces allow to see that even if the asymptotic behaviours (1), (2) and (3) seem to be close , they happen under hypotheses which are very different. In an appendix we will also state without proof some Kolmogorov-Brunk type SLLN which can also be obtained by applying the same Dubins-Freedman comparison result.

§ 2. PROHOROV'S BOUNDED LAW OF LARGE NUMBERS.

A sufficient condition for the BLLN in the classical Prohorov setting is as follows :

THEOREM 1 : Let (X_k) be a sequence of independent, centered, r.v. with values in a real separable p-uniformly smooth ($1 < p \leq 2$) Banach space $(B, \| \|)$, such that :

$$\exists K > 0 : \forall k \in \mathbb{N}, \ \|X_k\| \leq K (k / L_2 k) \quad \text{a.s.} \ ,$$

where : $L_2 x = \mathrm{Log}(\mathrm{Log} \ \sup (x, e))$.

Let's define for every integer n :

$$\Lambda(n) = 2^{-2n} \sum_{k \in I(n)} \sup (E f^2(X_k) : \|f\|_{B'} \leq 1) \quad ,$$

where : $I(n) = (2^n + 1, \ldots, 2^{n+1})$,

and suppose that the following hold :

a) The sequence (S_n / n) is stochastically bounded .

b) The sequence ($2^{-2np} \mathrm{Log}\, n \sum_{k \in I(n)} \|X_k\|^{2p}$) is stochastically bounded.

c) $\exists \ \varepsilon > 0 : \sum_{n \geq 1} \exp(-\varepsilon / \Lambda(n)) < +\infty$.

Then : $P(\sup_n \| S_n / n \| < +\infty) = 1$.

REMARK : One checks easily that condition b) above holds for instance if the sequence ($n^{-2p} L_2 n \sum_{1 \leq k \leq n} \|X_k\|^{2p}$) is stochastically bounded.

Now, let's give the proof of Theorem 1.

PROOF : An easy symmetrization argument, similar to the one used in the proof of Lemma 2.1 of [17] shows that it suffices to prove Theorem 1 for symmetrically distributed r.v. X_k . So we only consider that case. Another symmetrization argument shows that b) implies that the sequence ($2^{-2np} \mathrm{Log}\, n \sum_{k \in I(n)} \|X_k\|^{2p}$) is L^1-bounded.

For technical simplicity we make the following three assumptions - which aren't a loss of generality - :

i) $\exists K' > 0 : 486 K' (2p + 2C_p) \leq 1/8$ and :

$$\forall n \geq 2, \ \forall j \in I(n) , \quad \|X_j\| \leq K' 2^n / \mathrm{Log}\, n \quad \text{a.s.} \ ,$$

where of course C_p denotes the constant involved in the fundamental inequality recalled in Section 1.

ii) $\sum_{n \geq 1} \exp(-4K'^2 / \Lambda(n)) < +\infty$.

iii) $\sup_n 2^{-2np} \mathrm{Log}\, n \sum_{k \in I(n)} E \|X_k\|^{2p} \leq K'^2$.

Now, let's start the proof itself.

If one denotes by T_n the r.v. $2^{-n} \sum_{k \in I(n)} X_k$, an easy application of the Borel-

Cantelli lemma and of the symmetry of the X_k shows that the conclusion of Theorem 1 holds if the following is true :

$$\exists\, t>0 : \sum_{n\geq 1} P(\|T_n\|>t) <+\infty . \qquad (4)$$

By symmetry [1], hypothesis a) implies :

$$\sup_n E\|S_n/n\|^p <+\infty ,$$

and therefore also : $\qquad c = \sup_n E\|T_n\|^p <+\infty .$

So (4) will hold if we find $t'>0$ such that :

$$\sum_{n\geq 1} P(\|T_n\|^p - E\|T_n\|^p > t') <+\infty .$$

In order to find such a t', we begin by looking for good bounds for the quantities :

$$u_n = P(\|T_n\|^p - E\|T_n\|^p > 2x) ,$$

where $x \geq \sup(2, c)$ will be specified later.

Suppose that the integer $n \geq 2$ is fixed and denote by $Z_1, \dots, Z_{2^n}$ the r.v. $(X_j/2^n)_{j\in I(n)}$; we will also denote T_n by T and $\Lambda(n)$ by Λ. The symbol Σ will denote a sum taken on the set of integers $(1,\dots,2^n)$.

For finding a bound for u_n , we will consider two cases :

First case : $\operatorname{Log} n \leq K'^2/\Lambda$.

By adapting a well known trick of Yurinskii [21] , Ledoux [19] has noticed that a r.v. of the type $\|T\|^p - E\|T\|^p$ can be written as a martingale :

$$\|T\|^p - E\|T\|^p = \Sigma\, \eta_k ,$$

where : $\eta_k = E(\|T\|^p \mid \mathcal{F}_k) - E(\|T\|^p \mid \mathcal{F}_{k-1})$,

and : $\mathcal{F}_k = \sigma(Z_1,\dots,Z_k)$, $\mathcal{F}_0 = (\emptyset, \Omega)$.

For our needs this decomposition is not precise enough, so we will refine it by setting :

$$\forall k=1,\dots,2^n , \qquad A_k = (\|Z_1+\dots+Z_{k-1}+Z_{k+1}+\dots+Z_{2^n}\| > y) ,$$

$$B_k = A_k^c ,$$

where $y>0$ will be chosen later, and :

$$\alpha_k = E(\|T\|^p I_{A_k} \mid \mathcal{F}_k) - E(\|T\|^p I_{A_k} \mid \mathcal{F}_{k-1}) ,$$

$$\beta_k = E(\|T\|^p I_{B_k} \mid \mathcal{F}_k) - E(\|T\|^p I_{B_k} \mid \mathcal{F}_{k-1}) .$$

The following inequality obviously holds :

$$P(\|T\|^p - E\|T\|^p > 2x) \leq P(\Sigma\,\alpha_k > x) + P(\Sigma\,\beta_k > x) .$$

In a first step we will bound $P(\Sigma\,\alpha_k > x)$.

In order to do this we notice for beginning that $\left(\sum_{1 \le k \le j} \alpha_k , \mathfrak{F}_j \right)_{1 \le j \le 2^n}$ is a martingale to which we plan to apply the following comparison result of Dubins and Freedman [3] :

LEMMA 1 : Let $(S_k , \mathfrak{F}_k)$ be a real valued martingale and denote by (Y_k) its increments. If one defines for every $k = 1, \dots, n$:
$V_k = E(Y_k^2 \mid \mathfrak{F}_{k-1})$ - with $\mathfrak{F}_0 = (\emptyset, \Omega)$ - then :
$\forall\ a > 0.\ \forall\ b > 0,\ P(\exists\ j = 1, \dots, n : (Y_1 + \dots + Y_j) \ge a (V_1 + \dots + V_j) + b) \le 1/(1+ab)$.

For applying this lemma to our situation we need first a bound for the r.v. $|\alpha_k|$; this bound will be obtained by an easy change in the computations of the proof of Lemma 1 in [19] :

$$E(\|T\|^p I_{A_k} \mid \mathfrak{F}_k) \le E(\|T-Z_k\|^p I_{A_k} \mid \mathfrak{F}_k) + p\, E(| F(T-Z_k)(Z_k) I_{A_k} | \mid \mathfrak{F}_k) + C_p\, E(\|Z_k\|^p I_{A_k} \mid \mathfrak{F}_k) ,$$

and :

$$E(\|T\|^p I_{A_k} \mid \mathfrak{F}_{k-1}) \ge E(\|T-Z_k\|^p I_{A_k} \mid \mathfrak{F}_{k-1}) - p\, E(|F(T-Z_k)(Z_k) I_{A_k}| \mid \mathfrak{F}_{k-1}) - C_p\, E(\|Z_k\|^p I_{A_k} \mid \mathfrak{F}_{k-1}) .$$

From these two inequalities one deduces :

$$\alpha_k \le p\, E(| F(T-Z_k)(Z_k) I_{A_k} | \mid \mathfrak{F}_k) + C_p\, E(\|Z_k\|^p I_{A_k} \mid \mathfrak{F}_k) + p\, E(| F(T-Z_k)(Z_k) I_{A_k} | \mid \mathfrak{F}_{k-1}) + C_p\, E(\|Z_k\|^p I_{A_k} \mid \mathfrak{F}_{k-1}) .$$

The same computation can obviously be done for $-\alpha_k$, so :

$$|\alpha_k| \le p\, E(| F(T-Z_k)(Z_k) I_{A_k} | \mid \mathfrak{F}_k) + C_p\, E(\|Z_k\|^p I_{A_k} \mid \mathfrak{F}_k) + p\, E(| F(T-Z_k)(Z_k) I_{A_k} | \mid \mathfrak{F}_{k-1}) + C_p\, E(\|Z_k\|^p I_{A_k} \mid \mathfrak{F}_{k-1}) .$$

From this inequality it follows easily that :

$$E(\alpha_k^2 \mid \mathfrak{F}_{k-1}) \le 8p^2\, E(F^2(T-Z_k)(Z_k) I_{A_k} \mid \mathfrak{F}_{k-1}) + 8C_p^2\, E \|Z_k\|^{2p} E(I_{A_k} \mid \mathfrak{F}_{k-1}) .$$

If one puts now $\alpha = P(\bigcup_{1 \le k \le 2^n} A_k)$, one sees that :

$$\begin{aligned} E\ \Sigma\, E(\alpha_k^2 \mid \mathfrak{F}_{k-1}) &\le 8p^2 \sup_{1 \le k \le 2^n} E(\|T-Z_k\|^{2(p-1)} I_{A_k}) \Lambda + 8C_p^2 K'^2 \alpha / \mathrm{Log}\, n \\ &\le 8p^2\, \alpha^{3/4} \sup_{1 \le k \le 2^n} (E \|T-Z_k\|^{8(p-1)})^{1/4} \Lambda + 8C_p^2 K'^2 \alpha / \mathrm{Log}\, n \end{aligned}$$

By assumption a) and symmetry there exists $L > 0$, such that :

$$\sup_n \sup_{1 \le k \le 2^n} (E \|T-Z_k\|^{8(p-1)})^{1/4} \le L.$$

Hence :

$$E\,\Sigma\,E\,(\alpha_k^2 \mid \mathcal{F}_{k-1}) \le 8p^2 \wedge \alpha^{3/4} L + 8C_p^2 K'^2 \alpha / \mathrm{Log}\, n \quad ,$$

and so by assumption c) , one has for n large enough :

$$E\,\Sigma\,E\,(\alpha_k^2 \mid \mathcal{F}_{k-1}) \le \alpha^{3/4} \quad .$$

Applying now Lemma 1 for $a = \alpha^{-1/2}$ and $b = 1$, one obtains :

$$\begin{aligned} P(\Sigma\,\alpha_k > x) &\le P(\Sigma\,\alpha_k > x\,,\ \Sigma\,\alpha_k \ge \alpha^{-1/2}\,\Sigma\,E\,(\alpha_k^2 \mid \mathcal{F}_{k-1}) + 1) \\ &\quad + P(\Sigma\,\alpha_k > x\,,\ \Sigma\,\alpha_k < \alpha^{-1/2}\,\Sigma\,E\,(\alpha_k^2 \mid \mathcal{F}_{k-1}) + 1) \\ &\le \alpha^{1/2} + (\alpha^{1/4} / (x-1)) \le 2\,\alpha^{1/4} \quad . \end{aligned}$$

From the inequalities :

$$\alpha \le P(\sup_{1 \le k \le 2^n} \| \sum_{1 \le j \le k} Z_j \| > y/3) \le 2\,P(\| \Sigma\, Z_j \| > y/3) \quad ,$$

an application of Hoffmann-Jørgensen's Lemma ([14] Lemma 4.4) gives that for every $y \ge 81$:

$$\begin{aligned} P(\Sigma\,\alpha_k > x) &\le 2\,(2)^{3/4}\, P^{1/2}(\| \Sigma\, Z_k \| > y/9) \\ &\le 4\,(2)^{3/4}\, P(\| \Sigma\, Z_k \| > y/27) \le 32\, P^2(\| \Sigma\, Z_k \| > y/81). \end{aligned}$$

Now. fix $y \ge \sup(486,\ 243\,c)$, such that :

$$\forall n \in \mathbb{N}, \quad 32\, P(\| \Sigma\, Z_k \| > y/81) \le 1/2$$

- such a choice is of course possible by hypothesis a) - ; y being fixed, we put : $x = y/243$.

For such a couple (x, y) one has :

$$P(\| T \|^p - E\,\| T \|^p > 2x) \le 2\,P(\Sigma\,\beta_k > x) \tag{6}$$

In the next step of the proof we will bound the right-hand side of (6) by using another martingale result, also due to Dubins and Freedman [3] :

LEMMA 2 : <u>Let</u> $(S_k, \mathcal{F}_k)_{1 \le k \le n}$ <u>be a real valued martingale and denote by</u> (Y_k) <u>its increments</u>. <u>Suppose that</u> :

$$\forall k = 1, \ldots, n \qquad |Y_k| \le 1 \quad \text{a.s.}\ .$$

<u>Then</u> : $\forall u \in \mathbb{R}$, $\forall \lambda > 0$, $\forall v > 0$,

$$E\,\mathrm{ch}\left(\frac{\lambda(u + Y_1 + \ldots + Y_n)}{v + V_1 + \ldots + V_n} \right) \le \mathrm{ch}\,(\lambda u / v)\ \exp\,(v\, e(\lambda/v))\ ,$$

<u>where</u> : $\forall k = 1, \ldots, n$, $V_k = E\,(Y_k^2 \mid \mathcal{F}_{k-1})$ <u>and</u> :

$$e(x) = \exp x - 1 - x \quad .$$

For applying this lemma to our situation, we first notice that by the same computations as for the r.v. α_k one has :

$$| \beta_k | \leq p\, E(| F(T-Z_k)(Z_k)\, I_{B_k} | \mid \mathcal{F}_k) + C_p\, E(\| Z_k \|^p I_{B_k} \mid \mathcal{F}_k)$$
$$+ p\, E(| F(T-Z_k)(Z_k)\, I_{B_k} | \mid \mathcal{F}_{k-1}) + C_p\, E(\| Z_k \|^p I_{B_k} \mid \mathcal{F}_{k-1}) .$$

If for every $k = 1, \dots, 2^n$ one defines :

$$Y_k = (y^{1-p} \operatorname{Log} n \,/\, K'(2p+2C_p))\, \beta_k ,$$

one has : $\forall k = 1, \dots, 2^n$, $\quad |Y_k| \leq 1$ a. s. .

Now we want to bound :

$$a_n = P(\Sigma Y_k > x\, y^{1-p} \operatorname{Log} n \,/\, K'(2p+2C_p)) ,$$

by using Lemma 2.

Let's notice that :

$$V_1 + \dots + V_n \leq (\operatorname{Log} n)^2 (8p^2 \Lambda + 8C_p^2 K'^2 / \operatorname{Log} n) \,/\, K'^2 (2p+2C_p)^2 \leq 4 \operatorname{Log} n .$$

Hence :

$$a_n \leq P((1 / v+V_1+\dots+V_{2^n}) \Sigma Y_k > x y^{1-p} \operatorname{Log} n \,/\, K'(2p+2C_p) (v+4\operatorname{Log} n)) .$$

Remembering now that $p \in]1, 2]$ and also the assumption i) made on K', one obtains :

$$a_n \leq P((1 / v+V_1+\dots+V_{2^n}) \Sigma Y_k > 16 \operatorname{Log} n \,/\, (v+4\operatorname{Log} n)) .$$

Applying now Lemma 2 for $\lambda = v = 4 \operatorname{Log} n$, one gets :

$$a_n \leq \exp 4\operatorname{Log} n \,/\, \operatorname{ch}(8 \operatorname{Log} n) \leq 2\, n^{-4} .$$

So for the integers $n \geq n_0(x)$ and such that $\operatorname{Log} n \leq K'^2 / \Lambda$, one has :

$$u_n \leq 4\, n^{-4} .$$

Now we go to the second case.

<u>Second case</u> : $\operatorname{Log} n > K'^2 / \Lambda$.

It is easy to see that the relation (6) remains true - for the same couples (x, y) - and that the only thing to do is to apply Lemma 2 in a different manner.

If one defines :

$$\forall k = 1, \dots, 2^n, \qquad Y_k = (K' y^{1-p} \,/\, \Lambda (2p+2C_p))\, \beta_k ,$$

one has :

$$\forall k = 1, \dots, 2^n, \qquad | Y_k | \leq 1 \quad \text{a. s.} .$$

Furthermore :

$$V_1 + \dots + V_{2^n} \leq K'^2 (8p^2 + 8C_p^2 (K'^2 / \operatorname{Log} n)) \,/\, \Lambda^2 (2p+2C_p)^2 \leq 4 K'^2 / \Lambda .$$

So u_n can be bounded by :

$$2\, P((1 / v+V_1+\dots+V_{2^n}) \Sigma \beta_k > x K' y^{1-p} \,/\, (2p+2C_p) (\Lambda v + 4K'^2)) .$$

Now we apply Lemma 2 for $\lambda = v = 4K'^2 / \Lambda$ and we obtain :

$u_n \leq 2 \exp (4K'^2 / \Lambda) \; / \; \mathrm{ch} (8K'^2 / \Lambda) \leq 4 \exp - (4K'^2 / \Lambda)$.

Putting together the results obtained in the two cases, one gets :

$\forall n \geq N(x), \quad u_n \leq 4 (n^{-4} + \exp - (4K'^2 / \Lambda))$.

Hence :

$$\sum_{n \geq 1} u_n < +\infty ,$$

and this finishes the proof of Theorem 1.

§ 3. PROHOROV'S LAW OF LARGE NUMBERS IN THE WEAK TOPOLOGY.

It is natural to expect that this asymptotic behaviour can be obtained under some "weak topological refinement" of the hypotheses of Theorem 1. The precise result is as follows :

THEOREM 2 : Let (X_k) be a sequence of independent, centered, r. v. with values in a real separable, p-uniformly smooth ($1 < p \leq 2$) Banach space $(B, \| \|)$, such that :

$$\exists K > 0 : \forall k \in \mathbb{N}, \quad \| X_k \| \leq K (k / L_2 k) \quad \text{a. s.} \; .$$

Let's define for every integer n and every $f \in B'$:

$$\lambda(n, f) = 2^{-2n} \sum_{k \in I(n)} E f^2(X_k) \; .$$

Suppose that assumptions a), b), c) of Theorem 1 are fulfilled and that the following one holds also :

d) $\forall \varepsilon > 0 , \; \forall f \in B', \quad \sum_{n \geq 1} \exp (-\varepsilon / \lambda(n, f)) < +\infty$.

Then :

$$P(\omega : S_n(\omega) / n \to 0 \text{ weakly}) = 1.$$

PROOF : By Theorem 1, we know that :

$$P(\sup_n \| S_n / n \| < +\infty) = 1 \; .$$

A standard argument then gives [13] :

$$E \sup_n \| S_n / n \| < +\infty \; . \qquad (7)$$

Furthermore the one dimensional Prohorov SLLN [20] implies :

$$\forall f \in B' , \quad f(S_n / n) \to 0 \quad \text{a. s.} \; .$$

This property and (7) show that $(S_n / n , \sigma(X_1 , \dots , X_n))$ is a weak sequential amart of class (B) (for the definition and the main properties of weak sequential amarts, see for instance [4] V. 3).

The space B being reflexive, the conclusion of Theorem 2 follows immediately from a well known convergence theorem of weak sequential amarts due to Brunel and Sucheston [2].

§ 4. PROHOROV'S STRONG LAW OF LARGE NUMBERS.

The sufficient condition for the SLLN in the Prohorov setting can now be guessed easily from Theorems 1 and 2 ; the statement is as follows :

THEOREM 3 : Let (X_k) be a sequence of independent, centered, r.v. with values in a real, separable, p-uniformly smooth ($1 < p \leq 2$) Banach space $(B, \| \|)$, such that :

$$\exists K > 0 : \forall k \in \mathbb{N}, \quad \| X_k \| \leq K (k/L_2 k) \quad \text{a.s.}$$

Suppose that the following hold :

a') $S_n / n \to 0$ in probability .

b') $2^{-2np} \text{Log}\, n \sum_{k \in I(n)} \| X_k \|^{2p} \to 0$ in probability .

c') $\forall \varepsilon > 0, \quad \sum_{n \geq 1} \exp(-\varepsilon / \Lambda(n)) < +\infty$.

Then : $S_n / n \to 0$ a.s. .

REMARK : It is easy to see that b') holds for instance if :

$$n^{-2p} L_2 n \sum_{1 \leq k \leq n} \| X_k \|^{2p} \to 0 \text{ in probability .}$$

PROOF : By a classical argument [17] it is sufficient to consider the symmetrical case. By symmetrization (see [10] proof of Lemma 1) b') implies :

b'') $\lim_{n \to +\infty} 2^{-2np} \text{Log}\, n \sum_{k \in I(n)} E \| X_k \|^{2p} = 0.$

The ideas of the proof of Theorem 3 are the same as those used for proving Theorem 1 ; so we will keep the same notations as in the proof of Theorem 1 and we will only detail what needs to be detailed.
First one notices [1] that from a') it follows that : $E \| S_n / n \|^p \to 0$, and also : $E \| T_n \|^p \to 0$.

So the the conclusion of Theorem 3 will be true if we show :

$$\forall x > 0, \quad \sum_{n \geq 1} P(\| T_n \|^p - E \| T_n \|^p > 2x) < +\infty . \tag{8}$$

Without loss of generality it suffices to check (8) only for $x \in]0, 243^{-p}[$. Fix such an x.
As previously, we denote by u_n the general term of the series involved in (8) ; we will bound u_n by considering again two classes of indices n, slightly different

from those taken in the proof of Theorem 1.

Let's suppose - and this of course isn't a loss of generality - that :

$\forall n \geq 2 , \quad \forall j \in I(n) , \qquad \| X_j \| \leq 2^n / \mathrm{Log}\, n \qquad$ a. s. .

For simplicity we put $\theta = \exp (-8(2p+2C_p) /x)$.

We consider first the following situation :

<u>First case</u> : $\Lambda \,\mathrm{Log}\, n \leq \theta$.

It is easy to see that if one puts $y = 243\, x^{1/p}$, relation (6) holds for n large enough.

Now we will again apply Lemma 2 ; for this we introduce the following sequence (Y_k) :

$\forall k = 1, \ldots . 2^n \quad . \; Y_k = (\mathrm{Log}\, n/(2p+2C_p)) \; \beta_k$.

One has :

$\forall k = 1, \ldots . 2^n \quad . \; | Y_k | \leq 1$ a. s. ,

and :

$V_1 + \ldots + V_{2^n} \leq (2(\mathrm{Log}\, n)^2 \Lambda + 2\mathrm{Log}\, n \; d_n)$,

where : $d_n \to 0$, by condition b'') .

Therefore :

$\forall n \geq n(x) , \qquad V_1 + \ldots + V_{2^n} \leq 4\, \theta \,\mathrm{Log}\, n$.

Applying now Lemma 2 with $v = 4\,\theta\, \mathrm{Log}\, n$ and $\lambda = 4\, v\, (2p+2C_p) /x$, one gets :

$u_n \leq (2/\mathrm{ch} 2\mathrm{Log}\, n) \exp 4\, \theta^{1/2} \mathrm{Log}\, n \leq 4\, n^{-3/2}$.

It remains to consider the complementary situation :

<u>Second case</u> : $\Lambda \,\mathrm{Log}\, n > \theta$.

If one chooses now :

$$Y_k = (\theta / \Lambda (2p+2C_p)) \; \beta_k \quad ,$$

one has :

$\forall k = 1 \ldots . 2^n , \qquad | Y_k | \leq 1 \quad$ a. s. ,

and :

$\forall n \geq n'(x) . \qquad V_1 + \ldots + V_{2^n} \leq 4\, \theta^2 / \Lambda$.

Applying finally Lemma 2 for $v = 4\, \theta^2 / \Lambda$ and $\lambda = v\, x^2$ one gets :

$u_n \leq (2 / \mathrm{ch}(x^3 \theta^{7/8} / 2\Lambda)) \exp (4x^4\, \theta^2 / \Lambda) \leq 4 \exp (- x^3 \theta^{7/8} / 4\Lambda)$.

Collecting the partial results we obtain :

$\forall x \in]0, 243^{-p}[, \quad \exists N(x) \in \mathbb{N} , \quad \exists \gamma(x) > 0$:

$\forall n \geq N(x) , \quad u_n \leq 4 (n^{-3/2} + \exp (- \gamma(x) / \Lambda))$,

and (8) immediately follows from hypothesis c') .

REMARK : Hypothesis b') in Theorem 3 seems at first glance somewhat surprising and artificial. To shed light on its meaning we will now give some corollaries of Theorem 3 which will show that b') is a very weak hypothesis.

COROLLARY 1 : Let (X_k) be a sequence of independent, centered, r.v. with values in a real, separable, p-uniformly smooth ($1 < p \leq 2$) Banach space $(B, \|\ \|)$ $(B, \|\ \|)$, such that :

$\exists K > 0 : \forall k \in \mathbb{N}, \quad \| X_k \| \leq K\,(k/L_2 k) \quad$ a. s. .

Suppose that the following hold :

1) $n^{-p} \sum_{1 \leq k \leq n} \| X_k \|^p \to 0$ in probability.

2) $\forall\ \varepsilon > 0, \quad \sum_{n \geq 1} \exp(-\varepsilon / \Lambda(n)) < +\infty$.

Then : $\quad S_n / n \to 0$ a. s. .

The proof of this result is very easy. First :

$$2^{-2np} \sum_{k \in I(n)} \| X_k \|^{2p} \leq (c/\mathrm{Log}\, n)\, 2^{-2np} \sum_{k \in I(n)} \| X_k \|^p ,$$

and so hypothesis b') of Theorem 3 is fulfilled by applying 1).

By symmetrization one has :

$$n^{-p} \sum_{1 \leq k \leq n} E \| X_k \|^p \to 0 \quad ;$$

the space B being of type p, it follows that the sequence (S_n / n) converges to 0 in $L^p(B)$. All the assumptions of Theorem 3 being fulfilled, this ends the proof of Corollary 1.

In the case p = 2, Corollary 1 reduces to Prohorov's SLLN proved in [11] . Let's stay for a moment in this situation p = 2 for making precise the difference between the hypotheses of Theorem 3 and those of the result in [11] . One observes first that the following easy corollary of Theorem 3 holds :

COROLLARY 2 : Let (X_k) be a sequence of independent, centered, r.v. with values in a real, separable, 2-uniformly smooth Banach space $(B, \|\ \|)$, such that :

$\exists K > 0 : \forall k \in N . \ \| X_k \| \leq K\,(k/L_2 k) \quad$ a. s. ; .

Suppose that the following hold :

a') $S_n / n \to 0$ in probability .

b') $(1 / n^2 L_2 n) \sum_{1 \leq k \leq n} \| X_k \|^2 \to 0$ in probability .

c') $\forall\ \varepsilon > 0, \quad \sum_{n \geq 1} \exp(-\varepsilon / \Lambda(n)) < +\infty$.

Then : $\quad S_n / n \to 0$ a. s. .

The gap between hypothesis b') in the above statement and assumption 1) in Corollary 1 is clear. It is easy to give examples of sequences of r.v. which belong to the domain of application of Corollary 2, but not to the one of Corollary 1 ; the sequence considered in the last section of [18] provides such an example.

In [11] it has been noticed that in the special case of Hilbert spaces - which are of course 2-uniformly smooth - condition 1) in Corollary 1 is necessary for the SLLN in the Prohorov setting. More generally, is it possible to simplify the hypotheses of Theorem 3 by making additional cotype restrictions on B ?

Godbole has characterized the spaces of cotype q in terms of SLLN for the sequence ($\| X_k \|^q / k^{q-1}$) ([6] Theorem 2.1) ; unfortunately his result will of no help in our situation.

The (partial) result we are able to prove is as follows :

COROLLARY 3 : <u>Let</u> $3/2 \le p \le 2$ <u>and</u> $2 \le q \le 2p-1$; <u>consider a sequence</u> (X_k) <u>of independent, centered, r.v. with values in a real, separable, p-uniformly smooth Banach space</u> $(B, \| \|)$, <u>of cotype q.</u>

<u>Suppose that the following hold</u> :

i) $\exists K > 0 \ : \forall k \in N, \ \| X_k \| \le K (k/L_2 k)$ a.s. .

ii) $\forall \varepsilon > 0, \ \sum_{n \ge 1} \exp(-\varepsilon / \Lambda(n)) < +\infty$.

<u>Then</u> : $S_n / n \to 0$ a.s. $\Leftrightarrow$ $S_n / n \to 0$ <u>in probability</u> .

PROOF : Of course the only thing to do is to check that the weak law of large numbers implies the strong one. As $S_n / n \to 0$ in probability, it is sufficient to consider the symmetrical case.

By cotype q and assumption i), one has :

$$\lim_{n \to +\infty} n^{-q} \sum_{1 \le k \le n} E \| X_k \|^q = 0.$$

Hence :

$$n^{-2p} L_2 n \sum_{1 \le k \le n} E \| X_k \|^{2p} \le K^{2p-q} n^{-q} (L_2 n)^{q+1-2p} \sum_{1 \le k \le n} E \| X_k \|^q ,$$

and Corollary 3 immediately follows, by application of Theorem 3.

REMARK : Let's suppose that conditions i) and ii) of Corollary 3 are fulfilled for a sequence of symmetrically distributed r.v. X_k . If the weak law of large numbers holds, then, by Godbole's result $n^{-q} \sum_{1 \le k \le n} \| X_k \|^q \to 0$ a.s. . So the following implication holds :

$$S_n / n \to 0 \text{ in probability} \Rightarrow n^{-q} \sum_{1 \le k \le n} \| X_k \|^q \to 0 \text{ a.s.} ;$$

this shows the very special geometric nature of these spaces which are both p-uniformly smooth ($3/2 \le p \le 2$) and of cotype q ($2 \le q \le 2p-1$) .

§ 5. APPENDIX : SOME LAWS OF LARGE NUMBERS OF KOLMOGOROV-BRUNK TYPE.

In [9] and [12] the SLLN of Kolmogorov and Brunk are studied in 2-uniformly smooth spaces under hypotheses requiring both assumptions on the strong and on the weak moments of the r.v. . Results stronger than the classical ones known for type 2 spaces are obtained.

By applying - in a more elementary manner as above - the martingale technique developed in section 2, Kolmogorov-Brunk type SLLN can also be obtained in p-uniformly smooth ($1 < p \leq 2$) spaces ; in the case $p = 2$ these results are stronger than the ones given previously in [9] and [12] . The proofs being elementary, we only state the results.

THEOREM 4 : Let (X_k) be a sequence of independent, centered, r.v. with values in a real, separable, p-uniformly smooth ($1 < p \leq 2$) Banach space $(B, \| \|)$.
Suppose that :

i) There exists a sequence of positive numbers (d_j) , converging to 0, such that :

$$\forall j \in \mathbb{N}, \quad \| X_j \| \leq j\, d_j \quad \text{a. s.} \quad ,$$

ii) $S_n / n \to 0$ in probability .

If one defines for every integer n :

$$\Gamma(p, n) = 2^{-2np} \sum_{j \in I(n)} E \| X_j \|^{2p} \quad ,$$

then the following implication holds :

$$\left(\exists\, k \text{ integer, } k \geq 1 : \sum_{n \geq 1} (\Lambda(n) + \Gamma(p, n))^k < +\infty \right) \Rightarrow S_n / n \to 0 \text{ a. s.} \; .$$

As an obvious corollary of Theorem 4 one has the following result which can be compared with the well known law of large numbers in type p spaces of Hoffmann-Jørgensen and Pisier [15] :

COROLLARY 4 : Let (X_k) be a sequence of independent, centered, r.v. with values in a real, separable, p-uniformly smooth ($1 < p \leq 2$) Banach space $(B, \| \|)$.
Suppose that :

a) $S_n / n \to 0$ in probability .

b) $\sum_{k \geq 1} k^{-2p} E \| X_k \|^{2p} < +\infty$.

c) $\exists\, j \geq 1 : \sum_{n \geq 1} (\Lambda(n))^j < +\infty$.

Then : $\quad S_n / n \to 0 \quad$ a. s. .

Bounded laws of large numbers and laws of large numbers in the weak topology can of course also be considered in the Kolmogorov-Brunk setting. We only give an example of such a result :

THEOREM 5 : Let (X_k) be a sequence of independent, centered, r.v. with values in a real separable, p-uniformly smooth $(1 < p \leq 2)$ Banach space $(B, \|\ \|)$. Suppose that :

a) The sequence (S_n / n) is stochastically bounded.

b) $\sum_{k \geq 1} k^{-2p} E \| X_k \|^{2p} < +\infty$.

c) $\exists j \geq 1 : \sum_{n \geq 1} (\Lambda(n))^j < +\infty$.

Then : $P(\sup_n \| S_n / n \| < +\infty) = 1$.

Let's conclude by mentioning an application of the above results. It is well known that all the classical situations in which one can conclude that a Banach space valued, symmetrically distributed r.v. X - and it is always possible to reduce to that case - satisfies the LIL, are the union of a Prohorov type BLLN and of a SLLN ([16] , [7]) or of a law of large numbers in the weak topology [8] . Therefore it is not surprising that Ledoux's recent necessary and sufficient condition for the LIL in uniformly convex spaces [19] can also be obtained as a corollary of Theorem 1 and Theorem 5. The computations for proving this observation are left to the reader.

REFERENCES.

[1] DE ACOSTA, A. : Inequalities for B-valued random vectors with applications to the strong law of large numbers. Ann. Prob. 9 (1981), 157-161

[2] BRUNEL, A. et SUCHESTON, L. : Sur les amarts faibles à valeurs vectorielles. C.R. Acad. Sc. Paris 282, Sér. A (1976), 1011-1014

[3] DUBINS, L.E. and FREEDMAN, D.A. : A sharper form of the Borel-Cantelli lemma and the strong law. Ann. Math. Stat. 36 (1965), 800-807

[4] EGGHE, L. : Stopping time techniques for analysts and probabilists. Cambridge University Press - Cambridge 1984

[5] ENFLO, P. : Banach spaces which can be given an equivalent uniformly convex norm. Israel J. of Math. 13 (1972), 281-288

[6] GODBOLE, A. : Strong laws of large numbers and laws of the iterated logarithm in Banach spaces. PHD, Michigan State University 1984

[7] HEINKEL, B. : Relation entre théorème central-limite et loi du logarithme itéré dans les espaces de Banach. Z. Wahrscheinlichkeitstheorie verw. Gebiete 49 (1979), 211-220

[8] HEINKEL, B. : Sur la loi du logarithme itéré dans les espaces réflexifs. Séminaire de Probabilités 16 - 1980/81 - Lecture Notes in Math 920, 602-608

[9] HEINKEL, B. : On the law of large numbers in 2-uniformly smooth Banach spaces. Ann. Prob. 12 (1984), 851-857

[10] HEINKEL, B. : The non i.i.d. strong law of large numbers in 2-uniformly smooth Banach spaces. Probability Theory on Vector Spaces III - Lublin 1983 Lecture Notes in Math 1080, 90-118

[11] HEINKEL, B. : Une extension de la loi des grands nombres de Prohorov. Z. Wahrscheinlichkeitstheorie verw. Gebiete 67 (1984), 349-362

[12] HEINKEL, B. : On Brunk's law of large numbers in some type 2 spaces. to appear in "Probability in Banach spaces 5" - Medford 1984 - , Lecture Notes in Math

[13] HOFFMANN-JØRGENSEN, J. : Sums of independent Banach space valued random variables. Studia Math 52 (1974), 159-186

[14] HOFFMANN-JØRGENSEN, J. : Probability in Banach space. Ecole d'été de Probabilités de St Flour 6 - 1976 - Lecture Notes in Math 598, 1-186

[15] HOFFMANN-JØRGENSEN, J. and PISIER, G. : The law of large numbers and the central limit theorem in Banach spaces. Ann. Prob. 4 (1976), 587-599

[16] KUELBS, J. : Kolmogorov law of the iterated logarithm for Banach space valued random variables. Ill. J. of Math. 21 (1977), 784-800

[17] KUELBS, J. and ZINN, J. : Some stability results for vector valued random variables. Ann. Prob. 7 (1979), 75-84

[18] LEDOUX, M. : Sur une inégalité de H.P. Rosenthal et le théorème limite central dans les espaces de Banach. Israel J. of Math. 50 (1985), 290-318

[19] LEDOUX, M. : La loi du logarithme itéré dans les espaces de Banach uniformément convexes. C. R. Acad. Sc. Paris 300, Sér. I, n° 17 (1985), 613-616

[20] STOUT, W.F. : Almost sure convergence. Academic Press, New York 1974

[21] YURINSKII, V.V. : Exponential bounds for large deviations. Theor. Prob. Appl. 19 (1974), 154-155

ON THE SMALL BALLS CONDITION IN THE CENTRAL LIMIT THEOREM IN UNIFORMLY CONVEX SPACES

Michel Ledoux
Département de Mathématique, Université Louis-Pasteur
7, rue René-Descartes, F-67084 Strasbourg, France

Introduction. Let E be a Banach space. By random variable with values in E we mean a measurable map X from some probability space $(\Omega, \mathfrak{F}, \mathbb{P})$ into E equipped with its Borel σ-field $\mathcal{B}(E)$ such that the image of $\mathbb{P}$ by X defines a Radon probability measure on $\mathcal{B}(E)$. Denote by $(X_n)_{n \in \mathbb{N}}$ a sequence of independent copies of X and, for each $n \geq 1$, $S_n = X_1 + \dots + X_n$. A random variable X with values in E is said to satisfy the central limit theorem (CLT in short) if the sequence $(S_n/\sqrt{n})_{n \in \mathbb{N}}$ converges weakly to a Gaussian Radon probability measure on E.

In his remarkable work on the Glivenko-Cantelli problem, M. Talagrand [T] observed the following characterization of the CLT for random variables with a strong second moment which relates the central limit property to conditions on small balls : if X takes its values in E and $\mathbb{E}\{\|X\|^2\} < \infty$, then X satisfies the CLT iff, for each $\varepsilon > 0$,

$$\liminf_{n \to \infty} \mathbb{P}\{ \frac{\|S_n\|}{\sqrt{n}} < \varepsilon \} > 0 .$$

(Actually, as detailled in [T], this equivalence holds in the more general setting of non-separable range spaces and in the framework of empirical processes.)

Although it seems rather difficult to verify these conditions on small balls, the preceding property is intriguing since it reduces a central limit property in Banach spaces to some kind of weak convergence on the line by taking norm. This property also lies at some intermediate stage since, as we will see below, a random variable X with values in E satisfies the CLT iff there exists a compact set K in E such that

$$\liminf_{n \to \infty} \mathbb{P}\{ \frac{S_n}{\sqrt{n}} \in K \} > 0$$

and the sequence $(S_n/\sqrt{n})_{n \in \mathbb{N}}$ is stochastically bounded as soon as for some $M > 0$

$$\liminf_{n \to \infty} \mathbb{P}\{ \frac{\|S_n\|}{\sqrt{n}} < M \} > 0 .$$

M. Talagrand (oral communication) raised the question whether the equivalence he proved holds without the strong second moment assumption which is not necessary in general for the CLT . In this note, we answer this question in a positive way in uniformly convex spaces. Precisely, we will establish the following result :

THEOREM 1 . Let E be a uniformly convex Banach space and X a random variable with values in E . Then X satisfies the CLT iff

(i) $\lim_{t \to \infty} t^2 \, \mathbb{P}\{ \|X\| > t \} = 0$

and

(ii) for each $\varepsilon > 0$, $\liminf_{n \to \infty} \mathbb{P}\{ \frac{\|S_n\|}{\sqrt{n}} < \varepsilon \} > 0$.

This result will follow easily from a new quadratic estimate of sums of independent random variables in uniformly convex spaces obtained in [L2] .

Preliminary results. We begin this section by a characterization of the CLT which follows easily from the concentration's inequality of M. Kanter [K] . I am grateful to Prof. X. Fernique for useful informations on this result.

PROPOSITION 2 . Let X be a random variable with values in a Banach space E . Then X satisfies the CLT iff there is a compact set K in E such that

$$\liminf_{n \to \infty} \mathbb{P}\{ \frac{S_n}{\sqrt{n}} \in K \} > 0 . \qquad (1)$$

Further, the sequence $(S_n/\sqrt{n})_{n \in \mathbb{N}}$ is stochastically bounded iff for some $M > 0$

$$\liminf_{n \to \infty} \mathbb{P}\{ \frac{\|S_n\|}{\sqrt{n}} < M \} > 0 . \tag{2}$$

Proof. The necessity of (1) and (2) is obvious. Let us first show why (2) implies the stochastic boundedness of $(S_n/\sqrt{n})_{n \in \mathbb{N}}$. Assume to begin with that X is symmetric. There exist $\delta > 0$ and k_o such that for all integers $k \geq k_o$ and n

$$\mathbb{P}\{ \frac{\|S_{nk}\|}{\sqrt{nk}} < M \} > \delta .$$

By Kanter's inequality

$$\frac{3}{2} (1 + k\,\mathbb{P}\{ \frac{\|S_n\|}{\sqrt{n}} > M\sqrt{k} \})^{-\frac{1}{2}} \geq \delta$$

and thus

$$k\,\mathbb{P}\{ \frac{\|S_n\|}{\sqrt{n}} > M\sqrt{k} \} \leq \frac{9}{4\delta^2} - 1 .$$

It follows that the sequence $(S_n/\sqrt{n})_{n \in \mathbb{N}}$ is stochastically bounded and also that $\mathbb{E}\{\|X\|^{\alpha}\} < \infty$ for all $\alpha < 2$. In the non-symmetric case, let X' be an independent copy of X; the symmetric random variable $X - X'$ satisfies (2) (with $2M$ instead of M) and hence the preceding conclusions apply to $X - X'$. In particular, we have that $\mathbb{E}\{\|X\|\} < \infty$; therefore the strong law of large numbers combined with (2) implies that X must be centered. Hence the conclusion to the second part of Proposition 2 holds by classical considerations involving Jensen's inequality. The first part is established in the same way.

Since in cotype 2 spaces, random variables X such that $(S_n/\sqrt{n})_{n \in \mathbb{N}}$ is stochastically bounded satisfy the CLT [P-Z], the previous proposition yields immediately the following corollary.

COROLLARY 3 . Let E be a Banach space of cotype 2 and X a random variable with values in E. Then X satisfies the CLT iff (2) holds.

We now turn to the small balls condition. Since for a centered Gaussian Radon probability, each ball centered at the origin of positive radius has positive mass, it is clearly necessary for a random variable X to satisfy the CLT that

$$\text{for each } \varepsilon > 0 , \quad \liminf_{n \to \infty} \mathbb{P}\{ \frac{\|S_n\|}{\sqrt{n}} < \varepsilon \} > 0 . \tag{3}$$

M. Talagrand [T] showed that when $\mathbb{E}\{\|X\|^2\} < \infty$, (3) is also sufficient for X to satisfy the CLT . For the sake of completeness, we reproduce here Talagrand's proof of this result ; it will illustrate the idea we will use next in uniformly convex spaces.

THEOREM 4 . Let X be a random variable with values in a Banach space E such that $\mathbb{E}\{\|X\|^2\} < \infty$. Then X satisfies the CLT iff (3) holds.

Proof. Let $\varepsilon > 0$ be fixed and $\delta = \delta(\varepsilon) > 0$ be such that

$$\liminf_{n \to \infty} \mathbb{P}\{ \frac{\|S_n\|}{\sqrt{n}} < \varepsilon \} > \delta .$$

Choose a finite dimensional subspace H of E such that if T denotes the quotient map $E \to E/_H$ and $\|.\|$ the quotient norm given by $\|T(x)\| = d(x,H)$, then $\mathbb{E}\{\|T(X)\|^2\} \leq \delta.\varepsilon^2$. For each n , $\|T(S_n)\| - \mathbb{E}\{\|T(S_n)\|\}$ can be written as a martingale

$$\|T(S_n)\| - \mathbb{E}\{\|T(S_n)\|\} = \sum_{i=1}^{n} d_i$$

with increments d_i , $i = 1,\dots,n$, such that, for each i

$$\mathbb{E}\{d_i^2\} \leq \mathbb{E}\{\|T(X_i)\|^2\}$$

(cf [Y]) and thus by Chebyschev's inequality

$$\mathbb{P}\{ | \|T(\frac{S_n}{\sqrt{n}})\| - \mathbb{E}\{\|T(\frac{S_n}{\sqrt{n}})\|\} | > \varepsilon \} \leq \varepsilon^{-2} \mathbb{E}\{\|T(X)\|^2\} \leq \delta .$$

Since $\|T\| \leq 1$,

$$\liminf_{n \to \infty} \mathbb{P}\{ \|T(\frac{S_n}{\sqrt{n}})\| < \varepsilon \} > \delta ,$$

and hence, intersecting,

$$\limsup_{n \to \infty} \mathbb{E}\{\|T(\frac{S_n}{\sqrt{n}})\|\} < 2\varepsilon .$$

X therefore satisfies the CLT by classical arguments (cf [P2]).

A short analysis of this proof shows the central rôle of the martingale trick leading to the quadratic estimate and of the integrability condition $\mathbb{E}\{\|X\|^2\} < \infty$ providing tightness at some point. An improved version, in uniformly convex spaces, of the previous martingale argument of Yurinskii was recently obtained in [L2] in some work on the law of the iterated logarithm. It will allow us to establish Theorem 1 which thus characterizes in those spaces the CLT through the small balls condition (3) and a moment condition which is necessary for the CLT .

Recall that a Banach space E is uniformly convex if for each $\varepsilon > 0$ there is a $\delta = \delta(\varepsilon) > 0$ such that for all x,y in E with $\|x\| = \|y\| = 1$ and $\|x - y\| \geq \varepsilon$, one has $1 - \|\frac{x-y}{2}\| > \delta$. According to a well-known fundamental result of G. Pisier [P1] , every uniformly convex Banach space E is p-smooth for some $p > 1$ i.e. admits an equivalent norm (denoted again $\|.\|$) with corresponding modulus of smoothness

$$\rho(t) = \sup \{ \tfrac{1}{2}(\|x+ty\| + \|x-ty\|) - 1 , \ \|x\| = \|y\| = 1 \}$$

satisfying $\rho(t) \leq Kt^p$ for all $t > 0$ and some positive finite constant K . This p-smooth norm is uniformly Fréchet-differentiable away from the origin with derivative $D : E - \{0\} \to E^*$ such that if $F_p(x) = \|x\|^{p-1} D(x/\|x\|)$ for $x \neq 0$ and $F_p(0) = 0$, then $\|F_p(x)\| = \|x\|^{p-1}$ for all x in E and, for some constant $C > 0$ (cf [H-J]),

$$\|F_p(x) - F_p(y)\| \leq C \|x - y\|^{p-1} \qquad \text{for all } x,y \text{ in } E . \tag{4}$$

The following lemma was the key point in the proof of the main result of [L2] . It will allow to achieve our wish in the next section.

LEMMA 5 . Let E be a p-smooth Banach space for some $p > 1$ with norm $\|.\|$ satisfying (4) . Let also $(Y_i)_{i \leq n}$ be a finite sequence of independent bounded E-valued random variables and let $S = \sum_{i=1}^{n} Y_i$. Then $\|S\|^p - \mathbb{E}\{\|S\|^p\}$ can be written as a martingale

$$\|S\|^p - \mathbb{E}\{\|S\|^p\} = \sum_{i=1}^{n} d_i$$

with increments d_i , $i = 1,\ldots,n$, such that, for each i ,

$$\mathbb{E}\{d_i^2\} \leq 2p^2\, \mathbb{E}\{F_p^2(S - Y_i)(Y_i)\} + 2C^2\, \mathbb{E}\{\|Y_i\|^{2p}\}$$

where C is the constant appearing in (4) .

Before turning to the proof of Theorem 1 , let us point out that a quotient of a p-smooth Banach space E is also p-smooth, and, if $\|.\|$ denotes the p-smooth norm of E , property (4) holds true for any quotient norm of $\|.\|$ with uniform constant C .

Proof of Theorem 1 . We may and do assume that E is equipped with a p-smooth $\|.\|$ for some $p > 1$ for which (4) and Lemma 5 hold. By the previous remark, these will also hold for every quotient norm with uniform constant C . We assume moreover $p < 2$.

Condition (i) is well-known to be necessary for X to satisfy the CLT ([A-A-G] , [P-Z]). Let us show the suffiency part of the theorem and assume first that X is symmetric. Proposition 2 and (ii) imply that the sequence $(S_n/\sqrt{n})_{n \in \mathbb{N}}$ is stochastically bounded and thus, from the integrability results of [P2] , we know that

$$\sup_n \mathbb{E}\{(\frac{\|S_n\|}{\sqrt{n}})^p\} \leq K^p < \infty . \tag{5}$$

Let $\varepsilon > 0$ be fixed. For each n, define

$$u_i = u_i(n) = \frac{X_i}{\sqrt{n}} I_{\{\|X_i\| \le \sqrt{n}\}} \quad , \quad i = 1,\dots,n \quad ,$$

and set $U_n = \sum_{i=1}^{n} u_i$; (i) and (ii) combine to imply the existence of a real number $\delta = \delta(\varepsilon) > 0$ such that

$$\liminf_{n \to \infty} \mathbb{P}\{ \|U_n\| < \varepsilon \} > \delta . \tag{6}$$

Since the sequence $(S_n/\sqrt{n})_{n \in \mathbb{N}}$ is stochastically bounded under (ii) and E does not contain an isomorphic copy of c_0 , Theorem 5.1 of [P-Z] ensures that X is pregaussian, that is, there exists a Gaussian random variable G in E with the same covariance structure as X . The integrability of Gaussian random vectors allows then to choose a finite dimensional subspace H of E such that if T denotes the quotient map $E \to E/_H$,

$$\mathbb{E}\{\|T(G)\|^2\} \le \delta.\varepsilon^{2p} / 2p^2 K^{2(p-1)} . \tag{7}$$

We now apply Lemma 5 to the sum $T(U_n)$ in $E/_H$; F_p will therefore denote below the Fréchet derivative of the quotient norm of $E/_H$. For each n , we have by orthogonality,

$$\begin{aligned} &\mathbb{E}\{ | \|T(U_n)\|^p - \mathbb{E}\{\|T(U_n)\|^p\} |^2 \} \\ &\le 2p^2 \sum_{i=1}^{n} \mathbb{E}\{F_p^2(T(U_n - u_i))(T(u_i))\} + 2C^2 \sum_{i=1}^{n} \mathbb{E}\{\|T(u_i)\|^{2p}\} \\ &\le 2p^2 n^{-1} \mathbb{E}\{\|T(G)\|^2\} \sum_{i=1}^{n} \mathbb{E}\{\|U_n - u_i\|^{2(p-1)}\} + 2C^2 n \mathbb{E}\{\|u_1(n)\|^{2p}\} \end{aligned} \tag{8}$$

since by independence

$$\mathbb{E}\{F_p^2(T(U_n - u_i))(T(u_i))\} \le n^{-1} \sup_{x^* \in (E/_H)_1^*} \mathbb{E}\{|x^*(T(X))|^2\} \; \mathbb{E}\{\|T(U_n - u_i)\|^{2(p-1)}\}$$

(where the supremum runs over the unit ball $(E/_H)_1^*$ of the dual of $E/_H$) and

$$\sup_{x^* \in (E/_H)_1^*} \mathbb{E}\{|x^*(T(X))|^2\} = \sup_{x^* \in (E/_H)_1^*} \mathbb{E}\{|x^*(T(G))|^2\} \le \mathbb{E}\{\|T(G))\|^2\} .$$

Now, by symmetry and (5) , for each $i = 1,\dots,n$,

$$(\mathbb{E}\{\|U_n - u_i\|^{2(p-1)}\})^{1/2(p-1)} \leq (\mathbb{E}\{\|U_n\|^p\})^{1/p} \leq K .$$

Further,

$$n\,\mathbb{E}\{\|u_1(n)\|^{2p}\} \leq n^{1-p} \int_0^{\sqrt{n}} \mathbb{P}\{ \|X\| > t \}\, d\,t^{2p}$$

$$\leq n \int_0^1 \mathbb{P}\{ \|X\| > t\sqrt{n} \}\, d\,t^{2p}$$

so that

$$\lim_{n\to\infty} n\,\mathbb{E}\{\|u_1(n)\|^{2p}\} = 0$$

by (i) and dominated convergence. These observations and (8) therefore imply that

$$\limsup_{n\to\infty} \mathbb{P}\{ \,|\, \|T(U_n)\|^p - \mathbb{E}\{\|T(U_n)\|^p\} \,|\, > \varepsilon^p \}$$

$$\leq \varepsilon^{-2p} \left(2p^2 K^{2(p-1)}\, \mathbb{E}\{\|T(G)\|^2\} \right) \leq \delta$$

(by (7)) and since (6) holds, by intersection,

$$\limsup_{n\to\infty} \mathbb{E}\{\|T(U_n)\|^p\} < 2\,\varepsilon^p$$

which easily implies that X satisfies the CLT (using (i) one more time).

In the general case, let X' be an independent copy of X ; the symmetric random variable $X - X'$ satisfies (i) and (ii) and thus the CLT and therefore so does X .

Conclusion. It is an open problem to know whether Theorem 1 holds true in any Banach space ; since we used the fact that random variables satisfying (ii) are pregaussian in spaces which do not contain an isomorphic copy of c_0 , a general statement should probably include the condition

(iii) X is pregaussian

as an additional (necessary) assumption. It would also be interesting to know in what spaces, random variables satisfying (i) (and possibly (iii)) verify the CLT iff the sequence $(S_n/\sqrt{n})_{n\in\mathbb{N}}$ is stochastically bounded (which is weaker than (ii)). At the present, only trivial situations in which this happens (such that cotype 2 spaces or spaces satisfying Λ-Ros(2) [L1] like L_p $(1 \leq p < \infty)$ spaces) have been described.

References.

[A-A-G] de Acosta, A., Araujo, A., Giné, E. : On Poisson measures, Gaussian measures, and the central limit theorem in Banach spaces. Advances in Prob., vol. 4, 1-68, Dekker, New York (1978).

[H-J] Hoffmann-Jørgensen, J. : On the modulus of smoothness and the G_α-conditions in B-spaces. Aarhus Preprint Series 1974-75, n°2 (1975).

[K] Kanter, M. : Probability inequalities for convex sets. J. Multivariate Anal. 6, 222-236 (1978) .

[L1] Ledoux, M. : Sur une inégalité de H.P. Rosenthal et le théorème limite central dans les espaces de Banach. Israel J. Math. 50, 290-318 (1985).

[L2] Ledoux, M. : The law of the iterated logarithm in uniformly convex Banach spaces. Trans. Amer. Math. Soc. (May 1986).

[P1] Pisier, G. : Martingales with values in uniformly convex spaces. Israel J. Math. 20, 326-350 (1975).

[P2] Pisier, G. : Le théorème de la limite centrale et la loi du logarithme itéré dans les espaces de Banach. Séminaire Maurey-Schwartz 1975-76, exposés 3 et 4, Ecole Polytechnique, Paris (1976).

[P-Z] Pisier, G., Zinn, J. : On the limit theorems for random variables with values in the spaces L_p $(2 \leq p < \infty)$. Zeit. für Wahr. 41, 289-304 (1978).

[T] Talagrand, M. : The Glivenko-Cantelli problem. Annals of Math., to appear (1984).

[Y] Yurinskii, V.V. : Exponential bounds for large deviations. Theor. Probability Appl. 19, 154-155 (1974).

SOME REMARKS ON THE UNIFORM CONVERGENCE OF GAUSSIAN AND RADEMACHER FOURIER QUADRATIC FORMS

M. Ledoux
Département de Mathématique
Université Louis-Pasteur
67084 Strasbourg, France

M.B. Marcus*
Department of Mathematics
Texas A & M University
College Station, Texas 77843, U.S.A.

1. Introduction

Let $\{g_n\}_{n=0}^{\infty}$ be an i.i.d. sequence of normal random variables with mean zero and variance 1 and let $\{\varepsilon_n\}_{n=0}^{\infty}$ be a Rademacher sequence, i.e. a sequence of i.i.d. random variables where $P(\varepsilon_0 = 1) = P(\varepsilon_0 = -1) = \frac{1}{2}$. Let $\{a_{m,n}\}_{m,n=0}^{\infty}$ be a sequence of complex numbers satisfying $\sum_{m,n} |a_{m,n}|^2 < \infty$. We define Gaussian and Rademacher Fourier quadratic forms as follows :

$$X_g(s,t) = \sum_{m<n} a_{m,n} g_m g_n e^{i(ms+nt)} \quad , \quad (s,t) \in [0,2\pi]^2 \tag{1.1}$$

and

$$X_\varepsilon(s,t) = \sum_{m<n} a_{m,n} \varepsilon_m \varepsilon_n e^{i(ms+nt)} \quad , \quad (s,t) \in [0,2\pi]^2 \ . \tag{1.2}$$

We are concerned with the uniform convergence a.s. of the stochastic processes $X_g(s,t)$ and $X_\varepsilon(s,t)$. (By convergence we mean

$$\lim_{N\to\infty} \sum_{n=1}^{N} \sum_{m=0}^{n-1} a_{m,n} g_m g_n e^{i(ms+nt)}$$

and similarly for $X_\varepsilon(s,t)$.) It is known that if these series converge in probability then they converge uniformly a.s. and in L^p of their sup-norm. This

---

* Professor Marcus was a visiting Professor at the University of Strasbourg when this work was initiated. He is very grateful for the hospitality and friendship that was extended to him while he was there. His work on this paper was also supported by a grant from the National Science Foundation of the U.S.A..

is given in [2] and [3] respectively for Gaussian and Rademacher quadratic forms (cf. also [9]).

Let $\{g_n'\}$ and $\{\varepsilon_n'\}$ be independent copies of $\{g_n\}$ and $\{\varepsilon_n\}$. It follows from the decoupling inequalities that for all $p \geq 1$ and $X_g(s,t)$ and $X_\varepsilon(s,t)$ as in (1.1) and (1.2) with finitely many non zero $a_{m,n}$:

$$(1.3) \qquad E\|X_g(s,t)\|^p \sim E\|\sum_{m<n} a_{m,n} g_m g_n' e^{i(ms+nt)}\|^p$$

and

$$(1.4) \qquad E\|X_\varepsilon(s,t)\|^p \sim E\|\sum_{m<n} a_{m,n} \varepsilon_m \varepsilon_n' e^{i(ms+nt)}\|^p$$

where the constants of equivalence only depend on p and $\|.\|$ indicates $\sup_{(s,t)\in[0,2\pi]^2} |.|$. We will show how (1.3) and (1.4) follow from the usual decoupling inequalities in the Appendix.

We associate with $\{X_g(s,t)\}_{(s,t)\in[0,2\pi]^2}$ and $\{X_\varepsilon(s,t)\}_{(s,t)\in[0,2\pi]^2}$ the pseudometric

$$(1.5) \qquad d \equiv d((s,t),(s',t')) = \Big(\sum_{m<n} |a_{m,n}|^2 |e^{i(ms+nt)} - e^{i(ms'+nt')}|^2\Big)^{1/2}$$

where (s,t) , $(s',t') \in [0,2\pi]^2$. As usual let $N([0,2\pi]^2,d;\varepsilon)$ denote the metric entropy of $[0,2\pi]^2$ with respect to d , i.e. the minimum number of open balls of radius $\varepsilon > 0$ in the pseudometric d that covers $[0,2\pi]^2$. A number of people have recognized that

$$(1.6) \qquad J_1([0,2\pi]^2,d) = \int_0^\infty \log N([0,2\pi]^2,d;\varepsilon)\, d\varepsilon < \infty$$

is a sufficient condition for the uniform convergence a.s. of (1.1) or (1.2). It appears for example in [4] (stated in a somewhat stronger form involving majorizing measures). In fact C. Borell pointed this out to us and raised the question, can this sufficient condition be improved ? The suspicion that (1.6) can be improved is reasonable because there are some obvious cases in which

$$J_{1/2}([0,2\pi]^2,d) = \int_0^\infty (\log N([0,2\pi]^2,d;\varepsilon))^{1/2} d\varepsilon < \infty$$

is sufficient for the uniform convergence a.s. of $X_g(s,t)$ or $X_\varepsilon(s,t)$. One of these is when $a_{m,n}$ is a product ; (details will be given in § 2). Nevertheless, no smaller function of the metric entropy than the one appearing in (1.6) suffices as a sufficient condition for the uniform convergence a.s. of $X_g(s,t)$ or $X_\varepsilon(s,t)$. To be more precise, for all functions $f : \mathbb{R}^+ \to \mathbb{R}^+$ such that $f(0) = 0$, f is increasing and $\lim_{x\to\infty} f(x)/x = 0$ (and also satisfies a weak smoothness condition which will be given in § 2), we give examples of Gaussian and Rademacher Fourier quadratic forms which are not uniformly convergent a.s. but for which

$$(1.7)\qquad J_f([0,2\pi]^2,d) = \int_0^\infty f(\log N([0,2\pi]^2,d;\varepsilon))\,d\varepsilon < \infty .$$

These examples also show that contrary to our experience with Gaussian processes no condition on the metric d alone can give necessary and sufficient conditions for the uniform convergence a.s. of (1.1).

Pertaining to our study, X. Fernique [6] has obtained an important corollary of the deep recent result of M. Talagrand [13] on the existence of a majorizing measure for bounded Gaussian processes. Fernique's corollary deals with stationary vector valued Gaussian processes ; using this corollary (in a form adapted to our needs, Theorem 3.1 of this paper) and Theorem 1.1 of [8,Chapter I] we obtain the following theorem which extends the equivalence between Gaussian and Rademacher Fourier series to quadratic forms.

<u>Theorem 1.1</u>. Let $\{X_g(s,t)\}_{(s,t)\in[0,2\pi]^2}$ and $\{X_\varepsilon(s,t)\}_{(s,t)\in[0,2\pi]^2}$ be as given in (1.1) and (1.2). If only finitely many of the coefficients $a_{m,n}$ are non zero, then for all $p \geq 1$

$$(1.8)\qquad E\|X_g(s,t)\|^p \sim E\|X_\varepsilon(s,t)\|^p$$

where, as above, $\|\cdot\| = \sup_{(s,t)\in[0,2\pi]^2} |\cdot|$ and the constants of equivalence depend only on p. In particular, $X_g(s,t)$ converges uniformly a.s. iff $X_\varepsilon(s,t)$ does and (1.8) holds in this case.

Developping further the arguments of the proof of Theorem 1.1 , we characterize the a.s. uniform convergence of $X_g(s,t)$ and $X_\varepsilon(s,t)$ in terms of one-dimensional entropies following Fernique's characterization of boundedness and continuity of vector valued stationary Gaussian processes. Assume for notational convenience that $a_{m,n} = 0$ if $m \geq n$ and set for $s, s', t, t' \in [0,2\pi]$,

$$(1.9) \qquad d_1(s,s') = d((s,0),(s',0)) \quad \text{and} \quad d_2(t,t') = d((0,t),(0,t'))$$

and for $j = 1 , 2$

$$(1.10) \qquad J_{1/2}([0,2\pi],d_j) = \int_0^\infty (\log N([0,2\pi],d_j;\varepsilon))^{1/2} d\varepsilon \ .$$

Recall that the space $C_{a.s.}$, of a.s. continuous random Fourier series, is defined as the space of all sequences of complex numbers $a = \{a_n\}_{n=0}^\infty$ in ℓ^2 such that

$$\sum_n a_n g_n e^{int} \ , \quad t \in [0,2\pi]$$

converges uniformly a.s. equipped with the norm

$$[\![a]\!] = E \sup_{t \in [0,2\pi]} \left| \sum_n a_n g_n e^{int} \right| \ .$$

The dual space $C^*_{a.s.}$ of $C_{a.s.}$ has been carefully described by G. Pisier as a space of multipliers $M(2,\psi_2)$; we refer the reader to [10] , [8] for further details. We will not be directly concerned with this here although various interpretations of what we obtain can be given in the language of [10] .

Define for each m and n the one-dimensional sequences

$$(1.11) \qquad A_n^1 = \{a_{m,n}\}_{m=0}^\infty \quad \text{and} \quad A_m^2 = \{a_{m,n}\}_{n=0}^\infty \ .$$

Set further for each T in $C^*_{a.s.}$ and $j = 1 , 2$

$$d_j^T(t,t') = \left(\sum_n | < T , A_n^j > |^2 \, |e^{int} - e^{int'}|^2 \right)^{1/2} \ , \quad t, t' \in [0,2\pi]$$

whenever it is defined. As a corollary of the results of Talagrand and Fernique we obtain :

Theorem 1.2. Let $\{X_g(s,t)\}_{(s,t)\in[0,2\pi]^2}$ be as given in (1.1) with only finitely many non zero coefficients $a_{m,n}$. Then for $j = 1$ or 2

$$E\|X_g(s,t)\| \sim \Big(\sum_{m<n} |a_{m,n}|^2\Big)^{1/2} + J_{1/2}([0,2\pi],d_1) + J_{1/2}([0,2\pi],d_2) + \sup_{\substack{T\in C^*_{a.s.} \\ \|T\|\le 1}} J_{1/2}([0,2\pi],d_j^T) \tag{1.12}$$

where the constants of equivalence are numerical constants and $J_{1/2}([0,2\pi],d_j^T)$ is defined as in (1.10) with d_j^T instead of d_j.

We note that by considering Cauchy sequences Theorem 1.2 gives necessary and sufficient conditions for uniform convergence a.s. of $X_g(s,t)$ and, by Theorem 1.1, also of $X_\varepsilon(s,t)$. Actually it is easily seen that, by Theorem 1.1 of [8, Chapter I], (1.12) also implies (1.8).

It is customary to leave out the "diagonal" (i.e. the terms $a_{m,m}$) when studying random quadratics forms. For one thing (1.3) and (1.4) are no longer true, in general, if $a_{m,m} \neq 0$. Furthermore, the diagonal terms can be handled separately. We write

$$\sum_m a_{m,m} g_m^2 e^{im(s+t)} = \sum_m a_{m,m} e^{im(s+t)} + \sum_m a_{m,m}(g_m^2 - Eg_m^2)e^{im(s+t)} . \tag{1.13}$$

By standard symmetrization argument and Theorem 1.1 [8, Chapter I], the second series on the right in (1.13) converges uniformly a.s. if and only if

$$\sum_m a_{m,m} g_m e^{imu} , \quad u \in [0,2\pi]$$

converges uniformly a.s.. The first series on the right in (1.13) is a deterministic Fourier series and the dichotomy between unboundedness and uniform convergence a.s. is no longer relevant. Obviously in the Rademacher case one only gets $\sum_m a_{m,m} e^{im(s+t)}$ on the right in (1.13).

2. Random Fourier quadratic forms and entropy

The sufficient conditions that we know for the uniform convergence a.s. of (1.1) and (1.2) follow immediately from well known results and techniques. We will present them here for the convenience of the reader.

Theorem 2.1. If $J_1([0,2\pi]^2,d) < \infty$ the Fourier quadratic forms (1.1) and (1.2) converge uniformly a.s. and

$$(2.1) \qquad E \sup_{(s,t)\in[0,2\pi]^2} |X_g(s,t)| \le C\left(\left(\sum_{m<n} |a_{m,n}|^2\right)^{1/2} + J_1([0,2\pi]^2,d) \right)$$

where C is a numerical constant. (2.1) is also valid with X_g replaced by X_ε. Conversely, if either of the Fourier quadratic forms X_g or X_ε converge uniformly a.s. then $J_{1/2}([0,2\pi]^2,d) < \infty$.

Proof. Let $\{b_{m,n}\}$ be complex numbers satisfying $\sum_{m<n} |b_{m,n}|^2 < \infty$. One can easily obtain from the results of [12] or [14] that there exists an absolute constant $C > 0$ such that

$$(2.2) \qquad P\left(\left|\sum_{m<n} b_{m,n} g_m g_n\right| > \lambda \left(\sum_{m<n} |b_{m,n}|^2\right)^{1/2} \right) \le C \exp(-\lambda/C), \quad \forall \lambda > 0 .$$

An estimate similar to (2.2) holds when $\{g_n\}$ is replaced by $\{\varepsilon_n\}$, (see e.g. [1]). Using these estimates Theorem 2.1 follows from the usual extensions of Dudley's theorem as presented for example in [11] or [5] since we have that

$$E \exp \alpha \left| \frac{X_g(s,t) - X_g(s',t')}{d((s,t),(s',t'))} \right| < \infty$$

for some $\alpha > 0$ and similarly with X_g replaced by X_ε .

For the converse we note that by (1.3) the uniform convergence a.s. of (1.1) implies that of

$$\sum_m \left(\sum_n a_{m,n} g'_n \right) g_m e^{ims} , \quad s \in [0,2\pi]$$

in which, to simplify the notation, we take $a_{m,n} = 0$ if $m \geq n$. Since $\{g_m\}$ is independent from $\{\sum_n a_{m,n} g'_n\}$ and is symmetric, the uniform convergence a.s. of (1.1) implies, by Theorem 1.1 [8, Chapter I], that of

$$\sum_m \Big(\sum_n |a_{m,n}|^2 \Big)^{1/2} g_m e^{ims} \quad , \quad s \in [0,2\pi] \; ,$$

which is equivalent to

$$J_{1/2}([0,2\pi],d_1) < \infty \tag{2.3}$$

where d_1 is defined in (1.9). Similarly we see that

$$J_{1/2}([0,2\pi],d_2) < \infty \; . \tag{2.4}$$

By the triangle inequality the metric d defined in (1.5) satisfies

$$d((s,s'),(t,t')) \leq d_1(s,s') + d_2(t,t') \; .$$

Therefore

$$N([0,2\pi]^2,d;\varepsilon) \leq N([0,2\pi],d_1;\varepsilon/2)\, N([0,2\pi],d_2;\varepsilon/2) \tag{2.5}$$

and thus we see by (2.3), (2.4) and (2.5) that the uniform convergence a.s. of (1.1) implies $J_{1/2}([0,2\pi]^2,d) < \infty$. Exactly the same argument applies if (1.1) is replaced by (1.2).

In an analogous fashion one can show that

$$J_{r/2}([0,2\pi]^r,\tilde{d}) = \int_0^\infty (\log N([0,2\pi]^r,\tilde{d};\varepsilon))^{r/2} d\varepsilon < \infty$$

implies the uniform convergence a.s. of r-dimensional tetrahedral (i.e. attention is restricted to the subset of $a_{m_1,\ldots,m_r}$ for which $m_1 < m_2 < \ldots < m_r$) Gaussian and Rademacher Fourier forms. Here, as above, $\tilde{d}$ denotes the corresponding distance in L^2.

We now give examples which show that if one wishes to give metric entropy conditions for the uniform convergence a.s. of (1.1) or (1.2) solely in terms of the metric d, then the sufficient condition of Theorem 2.1 can not be improved.

Let $f : \mathbf{R}^+ \to \mathbf{R}^+$, $f(0) = 0$, be an increasing function satisfying $\lim_{x\to\infty} x/f(x) = \lim_{x\to\infty} f(x) = \infty$. Furthermore we assume that for some $C > 0$ and all n_0 large enough

$$(2.6) \qquad \sum_{n=n_0}^{\infty} \frac{f(2n)}{2^n} \le C \frac{f(2n_0)}{2^{n_0}} .$$

(Such a condition will follow if for some positive finite numbers M and K , $f(2x) \le K f(x)$ for all $x \ge M$.) Let $\{N_j\}$ be a sequence of integers such that $N_j > N_1 + \ldots + N_{j-1}$ and let $\{b_j\}$ be a sequence of positive numbers. Their precise values will be specified later but we assume already that $\sum_{j=1}^{\infty} b_j^2 N_j^2 < \infty$. Define

$$(2.7) \qquad X_g(s,t) = \sum_{j=1}^{\infty} b_j \sum_{\substack{m,n\in I(j)\\ m<n}} g_m g_n e^{i(ms+nt)} , \qquad (s,t) \in [0,2\pi]^2$$

where for each $j \ge 1$, $I(j) = \{ n : N_1 + \ldots + N_{j-1} \le n < N_1 + \ldots + N_j \}$, $N_0 = 0$. Note that $\# I(j) = N_j$. We will show that for some appropriate choice of $\{N_j\}$ and $\{b_j\}$ depending upon f , the process X_g defined by (2.7) does not converge uniformly a.s. even though

$$J_f([0,2\pi]^2,d) = \int_0^{\infty} f(\log N([0,2\pi]^2,d;\varepsilon))\, d\varepsilon < \infty ,$$

where d is the metric associated with X_g as given in (1.5).

By independence and Jensen's inequality we see that for each $J \ge 1$

$$E \Big\| \sum_{j=1}^{J} b_j \sum_{\substack{m,n\in I(j)\\ m<n}} g_m g_n e^{i(ms+nt)} \Big\| \ge \sup_{j\le J} b_j \, E \Big\| \sum_{\substack{m,n\in I(j)\\ m<n}} g_m g_n e^{i(ms+nt)} \Big\|$$

and, by Remark 1.3 [8,Chapter VI] , for each j

$$E \Big\| \sum_{\substack{m,n\in I(j)\\ m<n}} g_m g_n e^{i(ms+nt)} \Big\|$$

$$\ge \tfrac{1}{2} \Big(E \sup_{t\in[0,2\pi]} \Big| \sum_{n\in I(j)} g_n e^{int} \Big|^2 - \sqrt{2}\, N_j \Big)$$

$$\ge C\, N_j \log N_j .$$

(Throughout this example C will denote a strictly positive constant, possibly changing from line to line.) Thus we have that

$$(2.8)\qquad \sup_{J\geq 1} E\Big\| \sum_{j=1}^{J} b_j \sum_{\substack{m,n\in I(j)\\ m<n}} g_m g_n e^{i(ms+nt)}\Big\| \geq C \sup_{j\geq 1} b_j N_j \log N_j .$$

To obtain an upper bound for the left side of (2.8) we note that

$$d((s,s'),(t,t')) = \Big(\sum_{j=1}^{\infty} b_j^2 \sum_{\substack{m,n\in I(j)\\ m<n}} |e^{i(ms+nt)} - e^{i(ms'+nt')}|^2\Big)^{1/2}$$

$$\leq \delta(s,s') + \delta(t,t')$$

where δ denotes the one-dimensional metric given by

$$\delta(t,t') = \Big(\sum_{j=1}^{\infty} b_j^2 N_j \sum_{n\in I(j)} |e^{int} - e^{int'}|^2\Big)^{1/2} .$$

It follows that

$$J_f([0,2\pi]^2,d) \leq 2\int_0^{\infty} f(2\log N([0,2\pi],\delta;\varepsilon))\, d\varepsilon .$$

We now estimate this entropy. For $s>0$ and each j we define

$$\hat\sigma(s) = \sup_{|u|\leq s} \delta(u,0) , \qquad \hat\sigma_j(s) = \sup_{|u|\leq s} \Big(\sum_{n\in I(j)} |e^{inu} - 1|^2\Big)^{1/2} .$$

Following [8, p.123-124 and Lemma 3.6 Chapter II] we see that

$$\int_0^{\infty} f(2\log N([0,2\pi],\delta;\varepsilon))\, d\varepsilon$$

$$\leq C\Big(\Big(\sum_{j=1}^{\infty} b_j^2 N_j^2\Big)^{1/2} + \int_0^1 f\Big(2\log\frac{1}{s}\Big)\, d\hat\sigma(s)\Big)$$

$$\leq C\Big(\Big(\sum_{j=1}^{\infty} b_j^2 N_j^2\Big)^{1/2} + \sum_{j=1}^{\infty} b_j\sqrt{N_j}\int_0^1 f\Big(2\log\frac{1}{s}\Big)\, d\hat\sigma_j(s)\Big) .$$

Furthermore,

$$\int_0^1 f\Big(2\log\frac{1}{s}\Big)\, d\hat\sigma_j(s) \leq \sum_{k=0}^{\infty} f(2k+2)\,(\hat\sigma_j(2^{-k}) - \hat\sigma_j(2^{-(k+1)}))$$

$$\leq f(2)\,\hat\sigma_j(1) + \sum_{k=1}^{\infty} \hat\sigma_j(2^{-k})\,(f(2k+2) - f(2k)) .$$

Now for each j and $s>0$

$$(2.9)\qquad \hat\sigma_j(s) \leq \sup_{|u|\leq s}\Big(\sum_{n=0}^{N_j-1} |e^{inu} - 1|^2\Big)^{1/2} + \sup_{|u|\leq s} \sqrt{N_j}\, |e^{i(N_1+\ldots+N_{j-1})u} - 1|$$

Let us denote by $\hat{\tau}_j(s)$ the first term of the right hand side of (2.9). It is easy to see that $\hat{\tau}_j(s) \le 2\sqrt{N_j}$, however, when $N_j \le 2^k$,

$$\hat{\tau}_j(2^{-k}) \le 2^{-k}\Big(\sum_{n=0}^{N_j-1} n^2\Big)^{1/2} \le 2^{-k} N_j^{3/2} ,$$

for j sufficiently large. Using these estimates for $\hat{\tau}_j(2^{-k})$ and (2.6) we see that for j sufficiently large

$$\begin{aligned}\sum_{k=1}^{\infty} \hat{\tau}_j(2^{-k})\,(f(2k+2) - f(2k)) &\le \sum_{k=1}^{[\log_2 N_j]} 2\sqrt{N_j}\,(f(2k+2) - f(2k)) \\ &\quad + \sum_{k=[\log_2 N_j]+1}^{\infty} N_j^{3/2}\, 2^{-k}\,(f(2k+2) - f(2k)) \\ &\le C\sqrt{N_j}\, f(3\log N_j) .\end{aligned}$$

By a similar decomposition of the sum on k we get

$$\begin{aligned}&\sum_{k=1}^{\infty} \sqrt{N_j}\,[(N_1+\dots+N_{j-1})\,2^{-k} \wedge 2\,]\,(f(2k+2) - f(2k)) \\ &\quad\le \sum_{k=1}^{\infty} \sqrt{N_j}\,(N_j\,2^{-k} \wedge 2)\,(f(2k+2) - f(2k)) \le C\sqrt{N_j}\, f(3\log N_j) .\end{aligned}$$

Putting this all together we see that

$$J_f([0,2\pi]^2,d) \le C\Big(\Big(\sum_{j=1}^{\infty} b_j^2 N_j^2\Big)^{1/2} + \sum_{j=1}^{\infty} b_j N_j\, f(3\log N_j)\Big) . \tag{2.10}$$

Since f satisfies $\lim_{x\to\infty} f(x)/x = 0$, we can first choose $\{N_j\}$ such that

$$\lim_{j\to\infty} \frac{j^2 f(3\log N_j)}{\log N_j} = 0$$

and then take

$$b_j = (j^2 N_j f(3\log N_j))^{-1} , \quad j \ge 1 .$$

For these choices, we see from (2.10) that $J_f([0,2\pi]^2,d) < \infty$ even though, by (2.8) and the integrability properties of Gaussian quadratic forms, $X_g(s,t)$ does not converge uniformly a.s..

Thus (1.6) can not be replaced by (1.7), with any increasing function f satisfying $\lim_{x \to \infty} f(x)/x = 0$ and (2.6), as a sufficient condition for the uniform convergence a.s. of (1.1). Note that exactly the same argument applies for the corresponding Rademacher Fourier quadratic form, i.e., with $\{\varepsilon_m \varepsilon_n\}$ replacing $\{g_m g_n\}$ in (2.7).

In some cases $J_{1/2}([0,2\pi]^2,d) < \infty$ is necessary and sufficient for the uniform convergence a.s. of (1.1) and (1.2). One of these, which is trivial, is when the coefficients $\{a_{m,n}\}$ vanish outside some one-dimensional set of indices. We write (1.1) in the form

$$(2.11) \qquad \sum_k a_k g_{mk} g_{nk} e^{ik(ms+nt)} \quad , \qquad (s,t) \in [0,2\pi]^2$$

where $m < n$ are non-negative integers. Using Gaussian decoupling we see that this series converges uniformly a.s. if and only if

$$(2.12) \qquad \sum_k a_k g_{mk} g'_{nk} e^{ik(ms+nt)} \quad , \qquad (s,t) \in [0,2\pi]^2$$

converges uniformly a.s. where $\{g'_{nk}\}$ is an independent copy of $\{g_{nk}\}$. Since $\{g_{mk} g'_{nk}\}$ is an independent symmetric sequence in k it follows from Theorem 1.1 [8, Chapter I] that the series in (2.12) and consequently (2.11) converges uniformly a.s. if and only if $J_{1/2}([0,2\pi]^2,d) < \infty$. Once again exactly the same argument holds if $\{g_{mk} g_{nk}\}$ in (2.11) is replaced by $\{\varepsilon_{mk} \varepsilon_{nk}\}$.

The finiteness of $J_{1/2}([0,2\pi]^2,d)$ also implies the continuity a.s. of the marginal processes formed from $X_g(s,t)$ and $X_\varepsilon(s,t)$.

<u>Theorem 2.2</u>. Let $X_g(s,t)$ be as given in (1.1). Then if $J_{1/2}([0,2\pi]^2,d) < \infty$ the processes $X_g(s,t_o)$ and $X_g(s_o,t)$ are uniformly convergent a.s. for each fixed $t_o \in [0,2\pi]$ and $s_o \in [0,2\pi]$. Conversely if $X_g(s,t_o)$ and $X_g(s_o,t)$ converge uniformly a.s. for some $t_o \in [0,2\pi]$ and $s_o \in [0,2\pi]$ then $J_{1/2}([0,2\pi]^2,d) < \infty$. This theorem is also valid if $X_g(s,t)$ is replaced by $X_\varepsilon(s,t)$.

Proof. Let $d_1(s,s')$ and $d_2(t,t')$ be as defined in (1.9). Let us note that

$$N([0,2\pi],d_j;2\varepsilon) \le N([0,2\pi],d;\varepsilon) , \quad j = 1,2 . \tag{2.13}$$

We will show this for $j = 1$. The proof when $j = 2$ is completely similar. Let (s_k,t_k), $k = 1, \dots, N([0,2\pi]^2,d;\varepsilon)$ be the centers of the balls of radius ε that cover $[0,2\pi]^2$ in the metric d. For a fixed k consider

$$B_{\varepsilon,k} = \cup \{(s,0) \in [0,2\pi]^2 : d((s,0),(s_k,t_k)) < \varepsilon \} .$$

Let $(\tilde{s}_k,0)$ be some fixed element in $B_{\varepsilon,k}$. It follows from the triangle inequality that

$$B_{\varepsilon,k} \subset \cup \{(s,0) \in [0,2\pi]^2 : d((s,0),(\tilde{s}_k,0)) < 2\varepsilon \}$$

since

$$d((s,0),(\tilde{s}_k,0)) \le d((s,0),(s_k,t_k)) + d((s_k,t_k),(\tilde{s}_k,0)) .$$

Since $d_1(s,\tilde{s}_k) = d((s,0),(\tilde{s}_k,0))$ we have verified (2.13) when $j = 1$. It follows that if $J_{1/2}([0,2\pi]^2,d) < \infty$ then $J_{1/2}([0,2\pi],d_1) < \infty$ and this implies that

$$\sum_m \Big(\sum_n |a_{m,n}|^2 \Big)^{1/2} g_m e^{ims} , \quad s \in [0,2\pi] \tag{2.14}$$

converges uniformly a.s.. By Theorem 1.1 [8, Chapter I], the uniform convergence a.s. of (2.14) is equivalent to that of

$$\sum_m \Big(\sum_n a_{m,n} g'_n e^{int_o} \Big) g_m e^{ims} , \quad s \in [0,2\pi]$$

which, by decoupling inequalities similar to (1.3), implies the uniform convergence a.s. of $X_g(s,t_o)$. (To avoid confusing notation assume that $a_{m,n} = 0$ for all $m \ge n$ in all these series.) A similar argument shows that $X_g(s_o,t)$ converges uniformly a.s.. The converse follows from the proof of Theorem 2.1 since in the relevant part of the proof of Theorem 2.1 the only property of the continuity of $X_g(s,t)$ that is used is that its marginals are continuous. All the above statements remain valid when $X_g(s,t)$ is replaced by $X_\varepsilon(s,t)$.

The following corollary is immediate.

<u>Corollary 2.3</u>. Assume that $a_{m,n} = a_m a_n$ where $\{a_n\}$ is a real sequence in ℓ^2. Then $J_{1/2}([0,2\pi]^2,d) < \infty$ is necessary and sufficient for the uniform convergence a.s. of $X_g(s,t)$ and $X_\varepsilon(s,t)$.

<u>Proof</u>. Since $\{a_m^2\} \in \ell^1$, and because of (1.3), $X_g(s,t)$ converges uniformly a.s. if and only if $\sum_m a_m g_m e^{ims}$ converges uniformly a.s.. The result now follows from Theorem 2.2. The same proof is valid for $X_\varepsilon(s,t)$.

Now let us consider stochastic processes of the form

$$(2.15) \qquad \sum_{m<n} a_{m,n}\varepsilon_{m,n}g_m g_n e^{i(ms+nt)} \quad , \qquad (s,t) \in [0,2\pi]^2$$

where $\{a_{m,n}\}$ and $\{g_n\}$ are as given in (1.1) but where $\{\varepsilon_{m,n}\}$ is a doubly indexed i.i.d. sequence of random variables with $P(\varepsilon_{0,0} = 1) = P(\varepsilon_{0,0} = -1) = \frac{1}{2}$, which is independent of $\{g_n\}$. It follows from Theorem 1.1 [8, Chapter I] that the series in (2.15) converges uniformly a.s. if and only if $J_{1/2}([0,2\pi]^2,d) < \infty$ where d is as given in (1.5). This observation has an interesting interpretation. Let $(\Omega,\mathfrak{F},P)$ be a probability space on which $\{\varepsilon_{m,n}\}$ is defined. For each $\omega \in \Omega$ we consider the Gaussian Fourier quadratic form

$$(2.16) \qquad \sum_{m<n} a_{m,n}\varepsilon_{m,n}(\omega)g_m g_n e^{i(ms+nt)} \quad , \qquad (s,t) \in [0,2\pi]^2 .$$

The metric d (as defined in (1.5)) is the same for all these processes irregardless of $\omega \in \Omega$. Furthermore, if $J_{1/2}([0,2\pi]^2,d) < \infty$ then for almost all $\omega \in \Omega$ the series in (2.16) converge uniformly a.s. (with respect to the probability space supporting $\{g_n\}$). However, the series in (2.16) <u>do not necessarily converge uniformly a.s. for all $\omega \in \Omega$</u>. This can be seen from the examples we just gave of Gaussian Fourier quadratic forms which do not converge uniformly a.s. but for which $J_{1/2}([0,2\pi]^2,d) < \infty$. Thus, unlike the random Fourier series studied in [8] continuity of Gaussian Fourier quadratic forms is not determined by the L^2 metric d given in (1.5). Once again exactly the same argument applies if $\{g_n\}$ is replaced by $\{\varepsilon_n\}$ in (2.16).

Nevertheless, entropy still can be used to characterize a.s. uniform convergence of Gaussian and Rademacher Fourier quadratic forms. However, as we will see in the next section, it is necessary to consider the supremum of classical one-dimensional entropies over a family of metrics.

3. Gaussian random Fourier series with coefficients in a Banach space

Fernique's corollary [6] of Talagrand's theorem [13] is the following.

Theorem 3.1. Let $(B,\|.\|)$ be a Banach space with dual B^*. Let $\{x_n\}$ be a sequence of elements of B with only finitely many non zero terms. Then

$$(3.1)\qquad E \sup_{t\in[0,2\pi]} \|\sum_n g_n x_n e^{int}\| \sim E\|\sum_n g_n x_n\| + \sup_{\|x^*\|\le 1} E \sup_{t\in[0,2\pi]} |\sum_n g_n < x^*, x_n > e^{int}|$$

where $x^* \in B^*$.

Therefore uniform convergence a.s. of random Fourier series with coefficients in a Banach space is characterized through conditions involving a family of classical one-dimensional entropies. We will see in the proof of Theorem 1.2 that (3.1) can be used to obtain a similar result for Gaussian and Rademacher Fourier quadratic forms. Using a set of one-dimensional entropy conditions instead of a single two-dimensional entropy condition to characterize uniform convergence a.s. of random Fourier quadratic forms seems to be necessary, as was shown by the class of examples described in § 2. Theorem 3.1 sheds some light on these examples. Indeed, the quadratic form (2.7) can almost be realized as a random Fourier series with coefficients in a Hilbert space. Define a sequence $\{x_n\}$ of elements in ℓ^2 by setting

$$\forall j \ge 1, \quad \forall n \in I(j), \quad x_n = \left(\frac{b_j}{N_j}\right)^{1/2} \sum_{k\in I(j)} e_k$$

where $\{e_k\}$ denotes the canonical basis of ℓ^2. Consider

$$X(t) = \sum_n g_n x_n e^{int}, \quad t \in [0,2\pi].$$

Clearly

$$\|X(t)\|^2 = \sum_n \|x_n\|^2 + \sum_{m \neq n} \langle x_m, x_n \rangle g_m g_n e^{i(m-n)t}$$

(3.2)

$$= \sum_j b_j N_j + 2 \sum_j b_j \sum_{\substack{m,n \in I(j) \\ m<n}} g_m g_n \cos((m-n)t) .$$

Thus $\|X(t)\|^2$ is closely related to the quadratic form (2.7). For the choices of $\{b_j\}$ and $\{N_j\}$ in § 2, $X(t)$ is unbounded a.s. since for

$$x_j^* = \left(\frac{1}{N_j}\right)^{1/2} \sum_{k \in I(j)} e_k \qquad (\|x_j^*\| = 1)$$

we have

$$E \sup_{t \in [0,2\pi]} \left| \sum_n \langle x_j^*, x_n \rangle g_n e^{int} \right| = E \sup_{t \in [0,2\pi]} \left| b_j^{1/2} \sum_{n \in I(j)} g_n e^{int} \right|$$

$$\geq C \, (b_j N_j \log N_j)^{1/2}$$

for some absolute constant C . Likewise the examples of § 2 show that no condition of the form $J_f([0,2\pi],\delta) < \infty$ (see (1.7)) where

$$\delta(t,t') = \left(\sum_{m,n} |\langle x_m, x_n \rangle|^2 |e^{i(m-n)t} - e^{i(m-n)t'}|^2 \right)^{1/2}$$

is sufficient for the uniform convergence a.s. of the Gaussian quadratic form in (3.2). (The relationship between Theorem 3.1 and metric entropy conditions is clearly

$$E \sup_{t \in [0,2\pi]} \left\| \sum_n g_n x_n e^{int} \right\| \sim E \left\| \sum_n g_n x_n \right\| + \sup_{\|x^*\| \leq 1} J_{1/2}([0,2\pi], d_{x^*})$$

where

$$d_{x^*}(t,t') = \left(\sum_n |\langle x^*, x_n \rangle|^2 |e^{int} - e^{int'}|^2 \right)^{1/2} .)$$

We now use Theorem 3.1 to obtain Theorem 1.1.

Proof of Theorem 1.1. We will give the proof in the case $p = 1$. By (1.3) and (1.4) we will prove this theorem with $X_g(s,t)$ and $X_\varepsilon(s,t)$ replaced respectively by

$$X_g'(s,t) = \sum_{m<n} a_{m,n} g_m g_n' e^{i(ms+nt)} , \quad (s,t) \in [0,2\pi]^2$$

and

$$X'_\varepsilon(s,t) = \sum_{m<n} a_{m,n}\varepsilon_m\varepsilon'_n e^{i(ms+nt)} \quad , \quad (s,t) \in [0,2\pi]^2$$

where, to simplify the notation we take $a_{m,n} = 0$ if $m \geq n$. Recall that only finitely many of the $a_{m,n}$ are non zero. Let us define $\{g_m\}$ and $\{\varepsilon_m\}$ on the probability space $(\Omega,\mathfrak{F},P)$ and $\{g'_n\}$ and $\{\varepsilon'_n\}$ on the probability space $(\Omega',\mathfrak{F}',P')$ and denote their corresponding expectation operators by E_g, E_ε, $E_{g'}$ and $E_{\varepsilon'}$. Let E denote expectation on the product space $(\Omega\times\Omega',\mathfrak{F}\times\mathfrak{F}',P\times P')$. Let us fix $\omega \in \Omega$ and consider $\{\sum_m a_{m,n}g_m(\omega)e^{ims}\}$ as a sequence in n of elements in the Banach space of continuous complex valued functions on $[0,2\pi]$. By Theorem 3.1 we have

$$(3.3)\qquad \begin{aligned} E_{g'} \sup_{s,t} \Big|\sum_n \Big(\sum_m a_{m,n}g_m(\omega)e^{ims}\Big) g'_n e^{int}\Big| &\sim E_{g'}\sup_s \Big|\sum_n \Big(\sum_m a_{m,n}g_m(\omega)e^{ims}\Big) g'_n\Big| \\ &\quad + \sup_s E_{g'} \sup_t \Big|\sum_n \Big(\sum_m a_{m,n}g_m(\omega)e^{ims}\Big) g'_n e^{int}\Big| \end{aligned}$$

By Theorem 1.4 [8,Chapter I] we can replace $\{g'_n\}$ by $\{\varepsilon'_n\}$ in the last term in (3.3). Doing this and taking expectation with respect to $\{g_m\}$ we get

$$(3.4)\qquad \begin{aligned} E\sup_{s,t}|X'_g(s,t)| &\sim E\sup_s \Big|\sum_m \Big(\sum_n a_{m,n}g'_n\Big) g_m e^{ims}\Big| \\ &\quad + E_g \sup_s E_{\varepsilon'} \sup_t \Big|\sum_m \Big(\sum_n a_{m,n}\varepsilon'_n e^{int}\Big) g_m e^{ims}\Big| \ . \end{aligned}$$

Let us denote the first and second term to the right of the equivalence sign in (3.4) by I and II. By Theorem 1.4 [8,Chapter I] applied twice

$$\begin{aligned} I &\sim E\sup_s \Big|\sum_m \Big(\sum_n |a_{m,n}|^2\Big)^{1/2}\varepsilon_m e^{ims}\Big| \\ &\sim E\sup_s \Big|\sum_m \Big(\sum_n a_{m,n}\varepsilon'_n\Big)\varepsilon_m e^{ims}\Big| \leq E\sup_{s,t}|X'_\varepsilon(s,t)| \ . \end{aligned}$$

In analyzing II we repeat the steps of (3.3) and (3.4) and obtain

$$(3.5)\qquad \begin{aligned} II &\leq E\sup_{s,t} \Big|\sum_m \Big(\sum_n a_{m,n}\varepsilon'_n e^{int}\Big) g_m e^{ims}\Big| \\ &\sim E\sup_t \Big|\sum_m \Big(\sum_n a_{m,n}\varepsilon'_n e^{int}\Big) g_m\Big| + E_{\varepsilon'}\sup_t E_g \sup_s \Big|\sum_m \Big(\sum_n a_{m,n}\varepsilon'_n e^{int}\Big) g_m e^{ims}\Big| \end{aligned}$$

Denote the first term to the left of the equivalence sign in (3.5) by I' and the second by II'. Using the same argument as above we have

$$I' = E \sup_t \left| \sum_n \left(\sum_m a_{m,n} g_m \right) \varepsilon'_n e^{int} \right|$$

$$\sim E \sup_t \left| \sum_n \left(\sum_m a_{m,n} \varepsilon_m \right) \varepsilon'_n e^{int} \right| \le E \sup_{s,t} \left| X'_\varepsilon(s,t) \right| .$$

We use Theorem 1.4 [8, Chapter I] on II' and obtain

$$II' \sim E_{\varepsilon'} \sup_t E_\varepsilon \sup_s \left| \sum_m \left(\sum_n a_{m,n} \varepsilon'_n e^{int} \right) \varepsilon_m e^{ims} \right|$$

$$\le E \sup_{s,t} \left| X'_\varepsilon(s,t) \right| .$$

Thus we have shown that

$$E \|X'_g(s,t)\| \le C \, E \|X'_\varepsilon(s,t)\| \tag{3.6}$$

for some absolute constant C. The reverse inequality in (3.6) is obtained by the contraction principle. This completes the proof of Theorem 1.1.

We finally prove Theorem 1.2.

<u>Proof of Theorem 1.2</u>. It will be sufficient to show that

$$\begin{aligned} E \|X_g(s,t)\| \sim\; & E \sup_{s \in [0,2\pi]} \left| \sum_m \left(\sum_n |a_{m,n}|^2 \right)^{1/2} g_m e^{ims} \right| \\ & + E \sup_{t \in [0,2\pi]} \left| \sum_n \left(\sum_m |a_{m,n}|^2 \right)^{1/2} g_n e^{int} \right| \\ & + \sup_{\substack{T \in C^*_{a.s.} \\ \|T\| \le 1}} E \sup_{s \in [0,2\pi]} \left| \sum_m \langle T, A^j_m \rangle g_m e^{ims} \right| \end{aligned} \tag{3.7}$$

for $j = 1$ or 2 where as before $a_{m,n} = 0$ if $m \ge n$ and where the sequences $\{A^j_m\}$ have been defined in (1.11). Indeed, Theorem 1.4 [8, Chapter I] then clearly implies (1.12). Following the notation of the preceding proof we show (3.7) for $j = 2$ and with the left hand side replaced by $E \|X'_g(s,t)\|$ by (1.3). As in the proof of Theorem 1.1 (cf. (3.3), (3.4) and the estimate of I), we have

$$(3.8)\qquad \begin{aligned} E\sup_{s,t}|X'_g(s,t)| &\sim E\sup_s\Big|\sum_m\Big(\sum_n|a_{m,n}|^2\Big)^{1/2}g_m e^{ims}\Big| \\ &\quad + E_g\sup_s E_{g'}\sup_t\Big|\sum_n\Big(\sum_m a_{m,n}g_m e^{ims}\Big)g'_n e^{int}\Big| .\end{aligned}$$

Let us call II the second term to the right of (3.8). It is plain that

$$II = E_g \sup_s [\![\sum_m A_m^2 g_m e^{ims}]\!]$$

where $[\![\cdot]\!]$ is the norm on $C_{a.s.}$. (3.7) then follows easily from a new application of Theorem 3.1 but this time in $C_{a.s.}$. Indeed,

$$II \sim E_g [\![\sum_m A_m^2 g_m]\!] + \sup_{\substack{T\in C^*_{a.s.}\\ \|T\|\le 1}} E\sup_s\Big|\sum_m \langle T, A_m^2\rangle g_m e^{ims}\Big|$$

and by Theorem 1.4 [8,Chapter I] ,

$$\begin{aligned} E_g [\![\sum_m A_m^2 g_m]\!] &= E\sup_t\Big|\sum_n\Big(\sum_m a_{m,n}g_m\Big)g'_n e^{int}\Big| \\ &\sim E\sup_t\Big|\sum_n\Big(\sum_m |a_{m,n}|^2\Big)^{1/2} g'_n e^{int}\Big| .\end{aligned}$$

Appendix.

Proof of (1.3) and (1.4) : the argument that we give here was shown to us by Gilles Pisier. We will first prove (1.4). It follows from Theorem 2 , [7] that for $p \ge 1$

$$(A.1)\qquad E\|X_\varepsilon(s,t)\|^p \sim E\Big\|\sum_{m<n} a_{m,n}(\varepsilon_m\varepsilon'_n + \varepsilon'_m\varepsilon_n)e^{i(ms+nt)}\Big\|^p$$

where the constants of equivalence depend only on p . Therefore by the triangle inequality

$$E\|X_\varepsilon(s,t)\|^p \le C_p\, E\Big\|\sum_{m<n} a_{m,n}\varepsilon_m\varepsilon'_n e^{i(ms+nt)}\Big\|^p .$$

Therefore, taking (A.1) into consideration, we see that to establish (1.4) we need only show

$$(A.2)\qquad C'_p\, E\Big\|\sum_{m<n} a_{m,n}\varepsilon_m\varepsilon'_n e^{i(ms+nt)}\Big\|^p \le E\Big\|\sum_{m<n} a_{m,n}(\varepsilon_m\varepsilon'_n + \varepsilon'_m\varepsilon_n)e^{i(ms+nt)}\Big\|^p .$$

Let θ_0 be a real valued random variable uniformly distributed on $[0,2\pi]$ and let $\{\theta_n\}$ be an i.i.d. sequence. Let $\{\theta_n'\}$ be an independent copy of $\{\theta_n\}$. We define $\xi_n = e^{in\theta_n}$, $\xi_n' = e^{in\theta_n'}$. It follows from the contraction principle for quadratic forms, Theorem 1, [7] that (A.2) holds if

$$\text{(A.3)} \qquad E\|\sum_{m<n} a_{m,n}\xi_m\xi_n' e^{i(ms+nt)}\|^p \leq E\|\sum_{m<n} a_{m,n}(\xi_m\xi_n' + \xi_m'\xi_n)e^{i(ms+nt)}\|^p \equiv H .$$

We replace ξ_n by $\xi_n e^{ins_0}$, $\xi_n' = \xi_n' e^{int_0}$ and (s,t) by $(s-s_0, t-t_0)$ in the expression on the right in (A.3) where s_0 and t_0 are fixed in $[0,2\pi]$. These changes do not change the numerical value of this term, i.e.

$$H = E\|\sum_{m<n} a_{m,n}(e^{i(ms_0+nt_0)}\xi_m\xi_n' + e^{i(ns_0+mt_0)}\xi_m'\xi_n)e^{-i(ms_0+nt_0)}e^{i(ms+nt)}\|^p$$

$$= E\|\sum_{m<n} a_{m,n}(\xi_m\xi_n' + e^{i[(n-m)s_0 + (m-n)t_0]}\xi_m'\xi_n)e^{i(ms+nt)}\|^p .$$

Finally we note that by Jensen's inequality

$$H = \frac{1}{4\pi^2}\int_0^{2\pi}\int_0^{2\pi} H \, ds_0 dt_0$$

$$\geq E\|\sum_{m<n} a_{m,n}(\xi_m\xi_n' + \frac{1}{4\pi^2}\int_0^{2\pi}\int_0^{2\pi} e^{i[(n-m)s_0+(m-n)t_0]}ds_0 dt_0 \; \xi_m'\xi_n) \, e^{i(ms+nt)}\|^p$$

$$= E\|\sum_{m<n} a_{m,n}\xi_m\xi_n' e^{i(ms+nt)}\|^p .$$

Thus we have obtained (A.3) and consequently (1.4).

The same argument works in the Gaussian case. We introduce $\{\tilde{g}_n\}$ and $\{\tilde{g}_n'\}$ in place of $\{\xi_n\}$ and $\{\xi_n'\}$ in (A.3) where $\tilde{g}_n = g_n + ig_n'$ for $\{g_n\}$ and $\{g_n'\}$ as defined in the introduction and take $\{\tilde{g}_n'\}$ to be an independent copy of $\{\tilde{g}_n\}$. Since $\tilde{g}_n$ is rotationally invariant the exact same argument as above gives (1.3).

References

[1] A. Bonami : Etude des coefficients de Fourier des fonctions de $L^p(G)$. Ann. Inst. Fourier (Grenoble) 20 , p. 335-402 (1970).

[2] C. Borell : Tail probabilities in Gauss space. Vector space measures and applications, Dublin 1978. Lecture Notes in Math. 644 , p. 71-82 , Springer (1979).

[3] C. Borell : On the integrability of Banach space valued Walsh polynomials. Séminaire de Probabilités XIII. Lecture Notes in Math. 721 , p. 1-3 , Springer (1979).

[4] C. Borell : On polynomials chaos and integrability. Prob. and Math. Stat. 3 , p. 191-203 (1984).

[5] X. Fernique : Régularité de fonctions aléatoires non gaussiennes. Ecole d'été de St-Flour 1981. Lecture Notes in Math. 976 , p. 1-74 , Springer (1983).

[6] X. Fernique : Fonctions aléatoires gaussiennes à valeurs vectorielles. Preprint (1985).

[7] S. Kwapien : Decoupling inequalities for polynomial chaos. Preprint (1985).

[8] M.B. Marcus and G. Pisier : Random Fourier series with applications to harmonic analysis. Ann. Math. Studies 101 , Princeton Univ. Press (1981).

[9] G. Pisier : Les inégalités de Khintchine-Kahane d'après C. Borell. Séminaire sur la géométrie des espaces de Banach 1977-78, Ecole Polytechnique, Paris (1978).

[10] G. Pisier : Sur l'espace de Banach des séries de Fourier aléatoires presque sûrement continues. Séminaire sur la géométrie des espaces de Banach 1977-78, Ecole Polytechnique, Paris (1978).

[11] G. Pisier : Some applications of the metric entropy condition to harmonic analysis. Banach spaces, harmonic analysis and probability, Proceedings 1980-81. Lecture Notes in Math. 995 , p. 123-154 , Springer (1983).

[12] M. Schreiber : Fermeture en probabilité des chaos de Wiener. C.R. Acad. Sci. Paris, Série A , 265 , p. 859-862 (1967).

[13] M. Talagrand : Régularité des processus gaussiens. C.R. Acad. Sci. Paris, Série I , 301 , p. 379-381 (1985).

[14] D. Varberg : Convergence of quadratic forms in independent random variables. Ann. Math. Statist. 37 , p. 567-576 (1966).

RATES OF CONVERGENCE IN THE CENTRAL LIMIT THEOREM FOR EMPIRICAL PROCESSES

*by Pascal MASSART**

Université Paris-Sud
U.A. CNRS 743 "Statistique Appliquée"
Mathématiques, Bât. 425
91405 ORSAY (France)

SUMMARY :

In this paper we study the uniform behavior of the empirical brownian bridge over families of functions $\mathcal{F}$ bounded by a function F (the observations are independent with common distribution P). Under some suitable entropy conditions which were already used by Kolčinskii and Pollard, we prove exponential inequalities in the uniformly bounded case where F is a constant (the classical Kiefer's inequality (1961) is improved), as well as weak and strong invariance principles with rates of convergence in the case where F belongs to $L^{2+\delta}(P)$ with $\delta \in]0,1]$ (our results improve on Dudley, Philipp's results (1983) whenever $\mathcal{F}$ is a Vapnik-Červonenkis class in the uniformly bounded case and are new in the unbounded case).

Key words and phrases : Invariance principles, empirical processes, gaussian processes, exponential bounds.

1. INTRODUCTION

1.1. GENERALITIES.

Let $(X,\mathcal{X},P)$ be a probability space and $(x_n)_{n\geq 1}$ be some sequence of independent and identically distributed random variables with law P, defined on a rich enough probability space $(\Omega,\mathcal{A},Pr)$.

P_n stands for the empirical measure $\frac{1}{n}\sum_{i=1}^{n}\delta_{x_i}$ and we choose to call empirical brownian bridge relating to P the centered and normalized process $\nu_n = \sqrt{n}\,(P_n-P)$. Our purpose is to study the behavior of the empirical brownian bridge uniformly over $\mathcal{F}$, where $\mathcal{F}$ is some subset of $\mathcal{L}^2(P)$.

More precisely, we hope to generalize and sometimes to improve some classical results about the empirical distribution functions on $\mathbb{R}^d$ (here $\mathcal{F}$ is the collection of quadrants on $\mathbb{R}^d$), in the way opened by Vapnik, Červonenkis and Dudley .

In particular, the problem is to get bounds for:

(1.1.1) $\qquad Pr\,(\,\|\nu_n\|_{\mathcal{F}} > t)$, for any positive t ,

where $\|.\|_{\mathcal{F}}$ stands for the uniform norm over $\mathcal{F}$ and to build strong uniform approximations of ν_n by some regular gaussian process indexed by $\mathcal{F}$ with some speed of convergence, say (b_n) .

First let us recall the main known results about the subject in the classical case described above.

1.2. THE CLASSICAL BIBLIOGRAPHY.

We only submit here a succinct bibliography in order to allow an easy comparison with our results (for a more complete bibliography see [26]) .

Concerning the real case $(d=1)$, the results mentioned below do not depend on P and are optimal :

1.2.1.

(1.1.1) is bounded by $C\exp(-2t^2)$, where C is a universal constant according to Dvoretsky, Kiefer and Wolfowitz [24] ($C \leq 4\sqrt{2}$ according to [17]).

1.2.2.

The strong invariance principle holds with $b_n = \frac{\log(n)}{\sqrt{n}}$, according to Komlos, Major and Tusnady [37] .

In the multidimensional case $(d \geq 2)$.

1.2.3.

(1.1.1) is bounded by $C(\varepsilon) \exp(-(2-\varepsilon)t^2)$, for any $\varepsilon > 0$, according to Kiefer [34] . In this expression ε cannot be removed (see [35] but also [28]).

1.2.4.

The strong invariance principle holds with $b_n = n^{-\frac{1}{2(2d-1)}} \operatorname{Log}(n)$, according to Borisov [8].

This result is not known to be optimal, besides it can be improved when P is uniformly distributed on $[0,1]^d$. In this case we have :

1.2.5. If $d = 2$:

The strong invariance principle holds with $b_n = \frac{(\operatorname{Log}(n))^2}{\sqrt{n}}$, according to Tusnady [50].

1.2.6. If $d \geq 3$:

The strong invariance principle holds with $b_n = n^{-\frac{1}{2(d+1)}} (\operatorname{Log}(n))^{\frac{3}{2}}$, according to Csörgö and Révész [14] .

1.2.5 and 1.2.6 are not known to be optimal.

Let us note that even the asymptotic distribution of $\|\nu_n\|_F$ is not well known (the case where $d = 2$ and P is the uniform distribution on $[0,1]^2$ is studied in [12]).

Now we describe the way which has already been used to extend the above results.

1.3. THE WORKS OF VAPNIK, ČERVONENKIS, DUDLEY AND POLLARD.

Vapnik and Červonenkis introduce in [51] some classes of sets - which are generally called V.Č.-classes - for which they prove a strong Glivenko-Cantelli law of large numbers and an exponential bound for (1.1.1) .

P. Assouad studies these classes in detail and gives many examples in [3] (see also [40] for a table of examples).

The functional P-Donsker classes (that is to say those uniformly over which some central limit theorem holds) were introduced and characterized for the first time by Dudley in [20] and were studied by Dudley himself in [21] and later by Pollard in [44].

Some sufficient (and sometimes necessary, see [27] in case F is uniformly bounded) conditions for F to be a P-Donsker class used in these works are some kind of <u>entropy</u> conditions :

Conditions where functions are approximated from above and below (bracketing, see [20]) are used in case F is a P-Donsker class whenever P belongs to some restricted set restricted set of laws on X (P is often absolutely continuous with respect to the Lebesgue measure in the applications) whereas Kolčinskii and Pollard's conditions are used in case F is a P-Donsker class whenever P belongs to some set of laws including any finite support law (the V.Č.-classes are - under some measurability assumptions - the classes of sets of this kind, see [21]).

In our study we are interested in the latter kind of the above classes.

Let us recall the already existing results in this particular direction.

Whenever F is some V.Č.-class and under some measurability conditions, we have:

1.3.1.

(1.1.1) is bounded by $C(F,\varepsilon)\exp(-(2-\varepsilon)t^2)$ for any ε in $]0,1]$, according to Alexander in [1] and more precisely by :

$$C(F)\ (1+t^2)^{2048(D+1)}\ \exp(-2t^2)\quad,\quad \text{in [2]}^*\ ,$$

where D stands for the integer density of F (from Assouad's terminology in [3]) .

1.3.2.

(1.1.1) is bounded by $4e^8\left(\sum_{j=0}^{D}\binom{n^2}{j}\right)\exp(-2t^2)$, according to Devroye in [16].

---

* Our result of the same kind (inequality 3.3.1°)a) in the present work) seems to have been announced earlier (in [41]) than K. Alexander's one.

1.3.3.

The strong invariance principle holds with $b_n = n^{-\frac{1}{2700(D+1)}}$, according to Dudley and Philipp in [23].

Now let us describe the scope of our work more precisely.

2. ENTROPY AND MEASURABILITY.

From now on we assume the existence of a non-negative measurable function F such that $|f| \leq F$, for any f in $\mathcal{F}$.

We use in this work Kolčinskii's entropy notion following Pollard [44] and the same measurability condition as Dudley in [21]. Let us define Kolcinskii's entropy notion.

Let p be in $[1,+\infty[$. $A(X)$ stands for the set of laws with finite support and $P_F^{(p)}(X)$ for the set of the laws making F^p integrable.

2.1. DEFINITIONS.

Let ε be in $]0,1[$ and Q be in $P_F^{(p)}(X)$.

$N_F^{(p)}(\varepsilon,\mathcal{F},Q)$ stands for the maximal cardinality of a subset $\mathcal{G}$ of $\mathcal{F}$ for which:

$$Q(|f-g|^p) > \varepsilon^p Q(F^p)$$

holds for any f,g in $\mathcal{G}$ with $f \neq g$ (such a maximal cardinality family is called an ε-net of $(\mathcal{F},F)$ relating to Q). We set $N_F^{(p)}(.,\mathcal{F}) = \sup_{Q \in A(X)} N_F^{(p)}(.,\mathcal{F},Q)$.

$\text{Log}(N_F^{(p)}(.,\mathcal{F}))$ is called the (p)-entropy function of $(\mathcal{F},F)$.

The finite or infinite quantities :

$$d_F^{(p)}(\mathcal{F}) = \inf \{s>0 \ ; \ \limsup_{\varepsilon \to 0} \varepsilon^s N_F^{(p)}(\varepsilon,\mathcal{F}) < \infty\}$$

$$e_F^{(p)}(\mathcal{F}) = \inf \{s>0 \ ; \ \limsup_{\varepsilon \to 0} \varepsilon^s \text{Log}(N_F^{(p)}(\varepsilon,\mathcal{F})) < \infty\}$$

are respectively called (p)-entropy dimension and (p)-entropy exponent of $(\mathcal{F},F)$.

Entropy computations.

We can compute the entropy of F from that of a uniformly bounded family as follows :

Let $\mathcal{I} = \{\frac{f}{F} 1_{(F>0)}\ ,\ f \in \mathcal{F}\}$, then :

$$N_F^{(p)}(.,\mathcal{F}) \leq N_1^{(p)}(.,\mathcal{I})$$

For, given Q in $A(X)$, either $Q(F)=0$ and so $N_F^{(p)}(.,\mathcal{F},Q) = 1$, or $Q(F) > 0$, so $\frac{F^p}{Q(F^p)} \in A(X)$ and then :

$$N_F^{(p)}(.,F,Q) = N_1^{(p)}(.,\mathcal{I}, \frac{F^p}{Q(F^p)} Q) \leq N_1^{(p)}(.,\mathcal{I}) \quad .$$

Some other properties of the (p)-entropy are collected in [40].

The main examples of uniformly bounded classes with finite (p)-entropy dimension or exponent are described below.

2.2. COMPUTING A DIMENSION : THE V.Č.-CLASSES.

According to Dudley [20] on the one hand and to Assouad [3] on the other we have

$$d_1^{(p)}(\mathcal{S}) = pd$$

whenever $\mathcal{S}$ is some V.Č.-class with real density d (this notion can be found in [3]). Concerning V.Č.-classes of functions, an analogous computation and its applications are given in [45] . See also [21] for a converse.

2.3. COMPUTING AN EXPONENT : THE HÖLDERIAN FUNCTIONS.

Let d be an integer and α be some positive real number.

We write β for the greatest integer strictly less than α.

Whenever x belongs to $\mathbb{R}^d$ and k to $\mathbb{N}^d$, $|k|$ stands for $k_1+\ldots+k_d$ and D^k for the differential operator $\frac{\partial^{|k|}}{\partial x_1^{k_1}\ldots\partial x_d^{k_d}}$.

Let $||.||$ be some norm on $\mathbb{R}^d$.

Let $\Lambda_{\alpha,d}$ be the family of the restrictions to the unit cube of $\mathbb{R}^d$ of the β-differentiable functions f such that :

$$\max_{|k|\leq\beta} \sup_{x\in\mathbb{R}^d} |D^k f(x)| + \max_{|k|=\beta} \sup_{x\neq y} \frac{|D^k f(x)-D^k f(y)|}{||x-y||^{\alpha-\beta}} \leq 1 \quad .$$

Then, according to [36] on the one hand and using Dudley's arguments in [19] on the other, it is easy to see that :

$$e_1^{(p)} (\Lambda_{\alpha,d}) = \frac{d}{\alpha}$$

Measurability considerations.

Durst and Dudley give in [21] an example of a V.Č.-class S such that $||P_n-P||_S \equiv 1$.

So some measurability condition is needed to get any of the results we have in view.

So from now on we assume the following measurability condition (which is due to Dudley [21]) to be fulfilled :

(M) . $(X,\mathcal{X})$ is a Suslin space

. There exists some auxiliary Suslin space $(Y,\mathcal{Y})$ and some mapping T from Y onto F such that :

$$(x,y) \to T(y)(x) \quad \text{is measurable on } (X\times X, \mathcal{X}\otimes\mathcal{Y})$$

and we say that F is image admissible Suslin via (Y,T) .

This assumption is essentially used through one measurable selection theorem which is due to Sion [47] (more about Suslin spaces is given in [13]).

2.4. THEOREM.

Let H be some measurable subset of $X\times X$. We write A for its projection on X . Then A is universally measurable and there exists a universally measurable mapping from A to Y whose graph is included in H .

A trajectory space for brownian bridges.

We set :

$$l_T^\infty(F) = \{h : F\to\mathbb{R} \; ; \; h\circ T \text{ is bounded and measurable on } (Y,\mathcal{Y})\} .$$

We consider $l_T^\infty(F)$ as a measurable space equipped with the σ-field generated by the open balls relating to $\|\cdot\|_{F}$ (which is generally distinct from the Borel σ-field because $1_T^{\infty}(F)$ is not separable).

This trajectory space does not depend on P any more (as it was the case in [20]) but only on the measurable representation (Y,T) of F.

From now on for convenience we set :

$$(\Omega,A,Pr) = (X^{\infty}\times [0,1],\widehat{X^{\infty}} \otimes B([0,1]),\ \widehat{P^{\infty}} \otimes \lambda)$$

where λ stands for the Lebesgue measure on $[0,1]$, $B([0,1])$ for the Borel σ-field on $[0,1]$ and $(X^{\infty},\widehat{X^{\infty}},\widehat{P^{\infty}})$ for the completed probability space of the countable product $(X^{\infty},X^{\infty},P^{\infty})$ of copies of (X,X,P).

The following theorem points out how $1_T^{\infty}(F)$ is convenient as a trajectory space.

2.5. THEOREM.

For any a in $\mathbb{R}^n$, $\sum_{i=1}^{n} a_i\delta_{x_i}$ is measurable from Ω to $1_T^{\infty}(F)$.

Moreover, setting $U_b(F) = \{h : F \to \mathbb{R}$; h is uniformly continuous and bounded on $(F,\rho_P)\}$, $U_b(F)$ is included in $1_T^{\infty}(F)$. Provided that (F,ρ_P) is totally bounded this inclusion is measurable.

Where $L^2(P)$ is given the distance $\rho_P : (f,g) \to \sigma_P(f-g)$, with $\sigma_P^2 : f \to P(f^2)-(P(f))^2$. For a proof of 2.5. see [21] (sec. 9) and [40] where it is also shown that many reasonable families (in particular $\Lambda_{\alpha,d}$ and the "geometrical" V.C.-classes) fulfill (M).

2.6. REMARK.

Since F fulfills (M) it follows from [21] (sec. 12) that $\|P_n-P\|_F \to 0$ a.s. whenever $N_F^{(1)}(.,F) < \infty$ and therefore :

$$\sup_{Q\in P_F^{(2)}(X)} N_F^{(2)}(\varepsilon,F,Q) \le N_F^{(2)}(\tfrac{\varepsilon}{2},F) \quad \text{for any } \varepsilon \text{ in }]0,1[.$$

This implies that the local behavior of the entropy function is unchanged when taking the sup in 2.1. over the set of any reasonable law.

3. EXPONENTIAL BOUNDS FOR THE EMPIRICAL BROWNIAN BRIDGE

We assume in this section that for some constants u and v , $u \leq f \leq v$ for any f in $\mathcal{F}$; we set $U = v-u$ and $\mathcal{F}-u = \{f-u \, , \, f \in \mathcal{F}\}$.

The following entropy conditions are considered :

a) $d_U^{(2)} (\mathcal{F}-u) < \infty$

b) $e_U^{(2)} (\mathcal{F}-u) < 2$

Using a single method we build upper bounds for (1.1.1) that are effective in the following two situations :

1°) Observe that $\|\sigma_P^2\|_{\mathcal{F}} \leq \frac{U^2}{4}$; nothing more is known about the variance over $\mathcal{F}$. In this case we prove some inequalities which are analogous to Hoeffding's inequality [30].

2°) We assume that $\|\sigma_P^2\|_{\mathcal{F}} \leq \sigma^2$.

This time our inequalities are analogous to Bernstein's inequality (see Bennett [5]) .

3.1. DESCRIPTION OF THE METHOD.

We randomize from a sample which size is equal to $N=mn$. In Pollard's [44], Dudley's [20] or Vapnik and Červonenkis [51] symetrization technics, $m=2$ but here, following an idea from Devroye [16] , we choose a large m .

Effecting the change of central law : $P \to P_N$ with the help of a Paul Lévy's type inequality, we may study $\tilde{P}_n - P_N$ instead of $P_n - P$ where $\tilde{P}_n$ stands for the randomized empirical measure.

Choosing some sequence of - measurably selected - nets relating to P_N whose mesh decreases to zero and controlling the errors committed by passing from a net to another via some one dimensional exponential bounds, we can evaluate, conditionally to P_N , the quantity $\|\tilde{P}_n - P_N\|_{\mathcal{F}}$.

Randomization .

Setting $N=nm$ (m is an integer), let w be some random one-to-one mapping from $[1,n]$ into $[1,N]$ whose distribution is uniform (the "sample w is drawn without replacement").

The inequalities in the next two lemmas are fundamental for what follows :

3.1. LEMMA.

For any ξ in $\mathbb{R}^N$, we set $S_N = \sum_{i=1}^{N} \xi_i$, $\tilde{S}_n = \sum_{i=1}^{n} \xi_{w(i)}$,

$$\sigma_N^2 = \left(\frac{1}{N}\sum_{i=1}^{N} \xi_i^2\right) - \left(\frac{1}{N}\sum_{i=1}^{N} \xi_i\right)^2$$

and $U_N = (\max_{1\leq i\leq N} (\xi_i)) - (\min_{1\leq i\leq N} (\xi_i))$; the following three quantities are, for any positive ε , lower bounds for $- \operatorname{Log}\left(\frac{1}{2}\Pr\left(\left|\frac{\tilde{S}_n}{n} - \frac{S_N}{N}\right| > \varepsilon\right)\right)$:

1°) $\dfrac{2n\varepsilon^2}{U_N^2}$

2°) $\dfrac{n\varepsilon^2}{2\sigma_N^2 + \varepsilon U_N}$

3°) $\dfrac{n\varepsilon^2}{m\sigma_N^2}$

These bounds only depend on ξ through numerical parameters (U_N, σ_N).

Bound 3°) is new ; concerning 1°) (due to Hoeffding [30]) , Serfling's bound is better (see [46]) but brings no more efficiency when m is large.

The proof of lemma 3.1. is given in the appendix.

From now we write $\tilde{P}_n$ for the randomized empirical process $\frac{1}{n}\sum_{i=1}^{n} \delta_{X_{w(i)}}$. The inequality allowing us to study the randomized process rather than the initial one is the following :

3.2. LEMMA.

The random elements $\|\tilde{P}_n - P_N\|_F$ and $\|P_n - P\|_F$ are measurable.

Besides, whenever $\|\sigma_P^2\|_F \leq \rho^2$, the following holds :

$$(1 - \frac{\rho^2}{a^2\varepsilon^2 n'}) \ \Pr\left(\|P_n - P\|_F > \varepsilon\right) \leq \Pr\left(\|\tilde{P}_n - P_N\|_F > (1-a)\, \frac{n'}{N}\, \varepsilon\right)$$

for any positive ε and any a in $]0,1[$, where $n' = N-n$.

For a proof of this lemma see [16] using Dudley's measurability arguments in [21] (sec. 12) .

Statement of the results.

3.3. THEOREM.

The following quantities are, for any positive t and η , upper bounds for $\Pr\left(\|\nu_n\|_F > t\right)$:

1°) a) if $d_U^{(2)}\,(F-u) = 2d$,

$$0_{\eta,F}(1)\ (1 + \frac{t^2}{U^2})^{3(d+\eta)} \exp\left(-2\, \frac{t^2}{U^2}\right)$$

b) if $e_U^{(2)}\,(F-u) = \zeta < 2$,

$$0_{\eta,F}(1) \exp\left(0_{\eta,F}(1)\ (\tfrac{t}{U})^{k+\eta}\right) \exp\left(-2\, \frac{t^2}{U^2}\right)$$

where $k = \zeta(\frac{6-\zeta}{2+\zeta})$ (when ζ increases from 0 to 2 so does k).

2°) Suppose that $\|\sigma_P^2\|_F \leq \sigma^2$, with $\sigma \leq U$, then

a) if $d_U^{(2)}\,(F-u) = 2d$,

$$0_{\eta,F}(1)\ (\tfrac{\sigma}{U})^{-4(d+\eta)}\ (1 + \frac{t^2}{\sigma^2})^{3(d+\eta)} \exp\left(- \frac{t^2}{2(\sigma^2 + \frac{U}{\sqrt{n}}(3U+t))}\right)$$

b) if $e_U^{(2)}\,(F-u) = \zeta < 2$,

$$0_{\eta,F}(1) \exp\left(0_{\eta,F}(1)\ (\tfrac{\sigma}{U})^{-\zeta-\eta}\ (\tfrac{t}{\sigma})^{2p-\zeta+\eta} + 5(\tfrac{t}{\sigma})^{2p+\eta}\right) \exp\left(- \frac{t^2}{2(\sigma^2 + \frac{U}{\sqrt{n}}\ (3U\ (\frac{t}{\sigma})^{p+\eta} + t))}\right)$$

where $p = \frac{2\zeta(4-\zeta)}{4+\zeta(4-\zeta)}$ (when ζ increases from 0 to 2 so does 2p) .

The constants appearing in these bounds depend on F only through $N_U^{(2)}(.,F-u)$ and of course on η .

Comments.

- From section 2.2., the assumption $d_1^{(2)}(F) < \infty$ is typically fulfilled whenever F is some V.Č.-class with real density d.

Thus bound 1°) a) is sharper than those of 1.3.1. ; in another connection the factor $O(F,\eta)\ t^{2\eta d}$ in 1°) a) is specified in the appendix.

- In the classical case (i.e. F is the collection of quadrants on $\mathbb{R}^d$) , bound 1°) a) improves on 1.2.3. but is less sharp than 1.2.1. in the real case ; moreover the optimality of 1°) a) is discussed in the appendix where we prove that :

$$\lim_{n\to\infty} \Pr\left(\|\nu_n\|_F > t\right) \geq 2 \sum_{i=0}^{d-1} \cdot \frac{(2t^2)^i}{i!} \exp(-2t^2) \qquad (*)$$

- Suppose that $F = \Lambda_{\alpha,d}$ then, from section 2.3. we have $e_1^{(2)}(F) = \frac{d}{\alpha}$. In other respects, Bakhvalov proves in [4] that if P stands for the uniform distribution on $[0,1]^d$ then :

$$\|\nu_n\|_F \geq C\ n^{\frac{1}{2}-\frac{\alpha}{d}} \quad \text{surely}$$

Thus we cannot get any inequality of the 1°) or 2°) type in the situation where $e_1^{(2)}(F) > 2$.

The border line case :

For any modulus of continuity ϕ , we can introduce a family of functions $\Lambda_{\phi,d}$ in the same way as $\Lambda_{\alpha,d}$ by changing $u \to u^\alpha$ into ϕ and defining β as the greatest integer for which $\phi(u)\ u^{-\beta} \underset{u\to 0}{\to} 0$ holds.

It is an easy exercise, using Bakhvalov's method, to show that :

$$\|\nu_n\|_{\Lambda_{\alpha,d}} \geq C\ (\mathrm{Log}(n))^\gamma$$

provided that $\phi(u) = u^{\frac{d}{2}} (\log(u^{-1}))^\gamma$ and P is uniformly distributed on $[0,1]^d$. Of course $e_1^{(2)}(\Lambda_{\phi,d}) = 2$ and we cannot get bounds such as in theorem 3.3.

(*) So, there is a gap for the degree of the polynomial factor in the bound 3.3.2°)a) between $2(d-1)$ and $6(d+\eta)$.

But the above result is rather rough and we want to go further in the analysis of the families $\Lambda_{\phi,1}$ around the border line.

Then the (2)-entropy plays the same role for $\Lambda_{\alpha,1}$ concerning the Donsker property as the metric entropy in a Hilbert space for the Hilbert ellipsoids concerning the pregaussian property, that is to say that the following holds :

(i) $\Lambda_{\phi,1}$ is a functional P-Donsker class whenever $\int_0^1 (\mathrm{Log}(N_1^{(2)}(\varepsilon,\Lambda_{\phi,1})))^{\frac{1}{2}}\, d\varepsilon < \infty$.

(ii) $\Lambda_{\phi,1}$ is not a functional λ-Donsker class whenever $\phi(u) \underset{-}{\sim} \left(\frac{u}{|\mathrm{Log}(u)|}\right)^{\frac{1}{2}}$ (*)
and in this case we have $\mathrm{Log}(N_1^{(2)}(\varepsilon,\Lambda_{\phi,1})) \underset{-}{\sim} (\varepsilon\,\mathrm{Log}(\varepsilon))^{-2}$.

(i) follows from Pollard's central limit theorem in [44].

(ii) follows from a result of Kahane's in [32] about Rademacher trigonometric series. In fact, if we set $\phi(u) = \sqrt{\frac{u}{|\mathrm{Log}(u)|}}$, we have from [32] p. 66 that :
$t \to \sum_{n>1} \frac{\varepsilon_n e_n(t)}{Kn\mathrm{Log}(n)}$ belongs to $\Lambda_{\phi,1}$ with some probability $p_K \underset{K\to\infty}{\to} 1$, where (ε_n) is a Rademacher sequence and $e_n(t) = \sqrt{2}\cos(2\pi nt)$.

Let us consider a standard Wiener process on $L^2([0,1])$, we may write $(W(e_n))$ as $(\varepsilon_n|W(e_n)|)$ with (ε_n) being independent of $(|W(e_n)|)$.

So that, with probability more than p_K , the following holds :

$$\|W\|_{\Lambda_{\phi,1}} \geq \frac{1}{K} \cdot \sum_{n\geq 1} |W(e_n)| \frac{1}{n\mathrm{Log}(n)} \quad .$$

By the three series theorem the series $\sum |W(e_n)| \frac{1}{n\mathrm{Log}(n)}$ diverges to infinity almost surely and therefore W is almost surely unbounded on $\Lambda_{\phi,1}$.

The same property holds for any brownian bridge G for $f \to G(f) + fW(1)$ is some Wiener process provided that $W(1)$ is some $N(0,1)$ random variable independent of G . So $\Lambda_{\phi,1}$ is not pregaussian and (ii) is proved.

(*) We write $f \underset{-}{\sim} g$, when $0 < \underline{\lim}\,(fg^{-1}) \leq \overline{\lim}\,(fg^{-1}) < \infty$

"An upper bound in situation 2°) is also an oscillation control".

If we set $G_\sigma = \{f-g \; ; \; \sigma(f-g) < \sigma \; , \; f,g \in F\}$, it is not difficult to see that:

3.4. $$N^{(2)}_{2U}(.,G_\sigma+U) \leq (N^{(2)}_{U}(.,F-u))^2$$

thus changing U into 2U and d into 2d if necessary the upper bounds in situation 2°) hold with G_σ instead of F , the constants being independent of σ because of 3.4. .

In particular if F is a V.Č.-class with real density d, we set :

$$A(\sigma,n,t) = \Pr(\|\nu_n\|_{G_\sigma} > t) \quad .$$

At it is summarized in [23], Dudley shows in [20] that $A(\sigma,n,t) \leq t$ whenever t is small enough, $\sigma = 0\left(\frac{t}{|\mathrm{Log}(t)|}\right)$ and $n \geq 0(t^{-r})$ with $r > 8$.

Applying 3.3.2°) a) improves on this evaluation for then :
$A(\sigma,n,t) \leq t$ whenever t is small enough, $\sigma = 0\left(\frac{t}{|\mathrm{Log}(t)|}\right)$

$$\text{and} \quad n \geq 0\left(\left(\frac{t}{|\mathrm{Log}(t)|}\right)^{-4}\right) .$$

In order to specify in what way the constant in bound 2°)a) depends on F, we indicate the following variant of 3.3.2°)a) .

3.5. PROPOSITION.

If we assume that $N^{(2)}_U(\varepsilon,F-u) \leq C(\varepsilon_0\varepsilon)^{-2d}$ for any ε in $]0,1[$ and some ε_0 in $]0,1[$ that $\|\sigma^2_P\|_F \leq \sigma^2$ with σ not exceeding U, then there exists some ε_1 in $]0,1[$ depending only on ε_0 and a constant K depending only on C such that :

$$\Pr(\|\nu_n\|_F > t) \leq K\,\varepsilon_1^{-d}\left(\frac{\sigma}{U}\right)^{-4d}\left(1+\frac{t^2}{\sigma^2}\right)^{14d}\exp\left(-\frac{t^2}{2(\sigma^2+\frac{U(3U+t)}{\sqrt{n}})}\right)$$

From now on L stands for the function $x \to \mathrm{Log}(x \vee e)$.

3.6. COROLLARY.

Let (F_n) be some sequence of V.Č.-classes fulfilling (M) with entire densities (D_n) . Then (with the above notations) $\Pr(\|\nu_n\|_{G_{\sigma_n}} > t) \underset{n\to\infty}{\to} 0$ for any positive t ; whenever $\sigma_n^2 = o(1/(D_nL(D_n)))$ and $\sigma_n^{-2} = 0(\sqrt{n})$.

(Provided that $D_n = o(\frac{\sqrt{n}}{Ln})$, such a choice of σ_n does exist).

Comment.

According to Le Cam [38] (Lemma 2) and applying 3.6. the process $\{\nu_n(f), f \in F_n\}$ admits finite dimensional approximations whenever $D_n = o\,(\frac{\sqrt{n}}{Log(n)})$ and provided that Le Cam's assumption (A1) is fulfilled.

This result improves on Le Cam's corollary of proposition 3 where $D_n = O(n^{-\gamma})$ for some $\gamma < \frac{1}{2}$ is needed.

Proof of 3.6.

Let F be a V.Č.-class with entire density D and real density d .

Using Dudley's proof in [20] (more details are given in [40]) it is easy to show that, for any $w > d$ (or $w \geq d$ if d is "achieved"), we have :

$$N_1^{(2)}\ (\varepsilon,F) \leq K^{1+(1/2|Log\varepsilon|)}\ \exp\,(2w)\ (1 + 2|Log\varepsilon|)^w\ \varepsilon^{-2w}$$

for any ε in $]0,1[$, with in particular when $w=D$, $K = \frac{3}{2D!}\,(2D)^D$.

So from Stirling's formula we get :

$N_1^{(2)}\ (\varepsilon,F) \leq C_1^3\ e^{5D}\ 2^{3D}\ \varepsilon^{-4D}$ for any ε in $]0, \frac{1}{\sqrt{2}}]$ and some universal constant C_1 . Hence, for any ε in $]0,1[$ we have :

$$N_1^{(2)}\ (\varepsilon,F) \leq C_1^3\ (2e)^{5D}\ \varepsilon^{-4D}$$

thus, applying 3.5. to the class G_{σ_n} yields 3.6.

We propose below another variant of inequality 3.3.2°)a), providing an alternative proof of a classical result about the estimation of densities.

3.7. PROPOSITION.

If we assume that $d_U^{(2)}(F-u) = 2d < \infty$ and $\|\sigma_P^2\|_F \leq \sigma^2$ with $\frac{UV}{\sqrt{n}} \leq \sigma \leq U$ for some positive V , then there exists some positive constant C such that an upper bound of $Pr\ (\|\nu_n\|_F > t)$ is, for any positive t, given by :

$$O_{V,F,\eta}(1)\ (n(1 + \frac{t^2}{\sigma^2}))^{3(d+\eta)}\ (\frac{\sigma}{U})^{-4(d+\eta)}\ \exp\left(- \frac{t^2}{2\,(\sigma^2 + \frac{U}{\sqrt{n}}\ (CLLn\ (\frac{U}{\sqrt{n}} + \sigma)+t))}\right)$$

In the situation where U is large this inequality may be more efficient than 3.3.2°)a) .

Application to the estimation of densities : minimax risk.

Let K_M be the following kernel on $\mathbb{R}^k$:

$$K_M(y) = \psi(y'M_y) \quad \text{for any } y \text{ in } \mathbb{R}^k$$

where ψ is some continuous function with bounded variation from $\mathbb{R}$ into $[-\frac{1}{2},\frac{1}{2}]$ and M is some $k \times k$ matrix.

Pollard shows in [45] that the class

$$\mathcal{K} = \{K_M(.-x), \quad M \in \mathbb{R}^{k^2}, \ x \in \mathbb{R}^k\}$$

is a V.Č.-class of functions and so :

$N_1^{(2)}(\varepsilon,\mathcal{K}) \leq C\varepsilon^{-w}$ for any ε in $]0,1[$ where C and w depend only on k .

Now if we assume that P is absolutely continuous with respect to the Lebesgue measure on $\mathbb{R}^k$, the classical kernel estimator of its density f is :

$$\hat{f}_n(x) = h^{-k} P_n \left(K\left(\frac{\cdot - x}{h}\right)\right)$$

where K is a K_M with fixed ψ and M so that $\int K^2(x)dx < \infty$.

We set $\bar{f} = E(\hat{f}_n)$

Proposition 3.7. gives a control of the *random expression* $\hat{f}_n - \bar{f}$ by choosing :

$F = \{h^{-k/2}K(\frac{\cdot - x}{h}) \ , \ x \in \mathbb{R}^k\}$, $\sigma = C$ and $U = h^{-k/2}$ where $C^2 > \|f\|_\infty \int K^2(x)dx$.

So, if we assume that $\frac{n}{Ln} \geq h^{-k} \geq C^2$, we get, setting $D_n = \sup_x |\hat{f}_n(x) - f(x)|$:

$$\Pr(\sqrt{nh^k}\, D_n > t) \leq O(n^\alpha)\, t^\beta \exp - \frac{t^2}{2\left(C^2 + O\left(\frac{LLn}{\sqrt{nh^k}}\right) + \frac{t}{\sqrt{nh^k}}\right)}$$

for any t in $[1+\infty[$ and some positive α and β . Hence, after an integration:

$$E(\sqrt{nh^k}\, D_n) \leq T + O(n^\alpha)\, T^\beta \exp - \frac{T^2}{2\left(C^2 + O\left(\frac{LLn}{\sqrt{nh^k}}\right) + \frac{T}{\sqrt{nh^k}}\right)}$$

for any T in $[1, +\infty[$, provided that $nh^k \geq 4\beta^2$.

We choose $T = O(\sqrt{Ln})$, thus :

$$E(D_n) = O\left(\left(\frac{Ln}{nh^k}\right)^{\frac{1}{2}}\right)$$

Provided that f belongs to some subset of regular functions Θ , the *bias expression* $\bar{f}-f$ can be evaluated so that the minimax risk associated to the uniform distance on $\mathbb{R}^k$ and to Θ can be controlled with the same speed of convergence as in [29] , via an appropriate choice of h .

3.8. SKETCHES OF PROOFS OF 3.3., 3.5., 3.7. (More details are given in [42]) .

First, by studying the class $G = \{\frac{f-u}{U}, f \in F\}$ instead of F , we may assume that $u=0$ and $v=1$. Let us proof theorem 3.3.. All along the proof we need to introduce parameters such as : a, μ (in $]0,1[$) ; r, m (in $\mathbb{N}$) ; α (in $]1,+\infty[$); q (in $]0,2]$) and positive s,β and γ which are all chosen in due time.

We set : $N = mn$, $\varepsilon = \frac{t}{\sqrt{n}}$ and $\varepsilon' = (1 - \frac{1}{m})(1-a)\varepsilon$.

We write $Pr^{(N)}(.)$ for the probability distribution conditional on $(x_1,\dots,x_N)$ and $\|.\|$ instead of $\|.\|_F$ for short.

A bound for $Pr(\|\nu_n\| > t)$ will follow, via 3.2., from a bound for $Pr(\|\tilde{P}_n - P_N\| > \varepsilon')$ which is at first performed conditionally on $(x_1,\dots,x_N)$.

The chain argument.

Let $(\tau_j)_{j\geq 1}$ be a positive sequence decreasing to zero.

For each integer j a τ_j-net F_j can be measurably selected (with the help of 2.4). A projection π_j may be defined from F onto F_j so that $P_N((\pi_j f-f)^2) \leq \tau_j^2$ holds.

Then

$$\|(\tilde{P}_n - P_N) \circ (Id - \pi_r)\| \leq \sum_{j\geq r+1} \|(\tilde{P}_n - P_N) \circ (\pi_j - \pi_{j-1})\|$$

So, if (η_j) is a positive series such that $\sum_{j\geq r+1} \eta_j \leq \mu$ we get :

$$Pr^{(N)}(\|\tilde{P}_n-P_N\| > \varepsilon') \leq A + B$$

where A and B are the $(x_1,\dots,x_N)$-measurable variables :

$$A = N_r \; \|Pr^{(N)}(|(\tilde{P}_n-P_N) \circ \pi_r| > (1-\mu)\varepsilon')\| ,$$

$$B = \sum_{j\geq r+1} N_j^2 \; \|Pr^{(N)}(|(\tilde{P}_n-P_N) \circ (\pi_j-\pi_{j-1})| > \eta_j \varepsilon')\|$$

(where N_j stands for $N_1^{(2)}(\tau_j,F)$) .

A is the principal part of the above bound and B is the sum of the error terms.

Inequalities 1°) or 2°) of Lemma 3.1. are needed to control A according to whether case 1°) or 2°) is investigated.

Bound 3°) in Lemma 3.1. is used to control B , giving :

$$B \leq 2 \sum_{j \geq r+1} N_j^2 \exp\left(- \frac{n\, \varepsilon'^2 \eta_j^2}{4m\, \tau_{j-1}^2}\right) \qquad \textit{3.8.1.}$$

Choosing $\eta_j = (j-1)^{-\alpha}$ and $r = 2 + \left[\left(\frac{1}{\mu}\right)^{\frac{1}{\alpha-1}}\right]$, (so $\sum_{j>r+1} \eta_j \leq \mu$ holds whenever $\alpha \geq 2$), the control of the tail of series 3.8.1. is performed via the following elementary lemma :

3.8.2. Lemma.

Let $\psi : [r, +\infty[\to \mathbb{R}$. Provided that ψ is an increasing convex function, the following inequality holds :

$$\sum_{j \geq r+1} \exp(-\psi(j)) \leq \frac{1}{\psi'_d(r)} \exp(-\psi(r))$$

where ψ'_d stands for the right-derivative of ψ .

We choose $\beta = 1$ under assumption a) and $\beta = \left(\frac{\zeta}{2-\zeta}\right)$ under assumption b) .

Proof of theorem 3.3. in case 1°).

We choose $a = t^{-2}$, $m = [t^2]$ and $\tau_j = \frac{1}{\sqrt{n}} j^{-(\alpha+\beta)}$ and apply 3.1.1°), then :

$$A \leq 2\, N_r\, e^{10} \exp(-2t^2 (1-2\mu)) \qquad P^{\otimes N}\text{-a.s.}$$

Under assumption a).

Considering the type of inequality we are dealing with we may assume that $N_j \leq C\, t^{2d}\, j^{2(\alpha+1)d}$ (instead of $N_j \leq C'\, t^{2d'}\, j^{2(\alpha+1)d'}$ for any $d' > d$) .

We choose , $\mu = t^{-2}$ and $\alpha = \text{Max}\left(2,\, 1 + \frac{4d}{3\eta}\right)$, so :

$$A \leq O_{\eta,F}(1)\, (1+t^2)^{3(d+\eta)} \exp(-2\, t^2) \qquad P^{\otimes N}\text{-a.s.}$$

and

$$B = O_{\eta,F}(1) \exp(-2t^2) \qquad P^{\otimes N}\text{-a.s.}$$

whenever $t^2 \geq 7+4d(\alpha+1)$. Now the above estimates are deterministic, so using Lemma 3.2., theorem 3.3. is proved in situation 1°)a) .

With the idea of proving proposition 3.5. note that, setting $\alpha = 2$, the above method gives , under the hypothesis in 3.5. , that $\Pr(\|\nu_n\| > t)$ is bounded by :

$$K_1 (\varepsilon_0^{-1} t)^{4d} (2+t^2)^{12d} \exp(-2t^2)$$

with K_1 depending only on C , whenever $t^2 \geq 7+12d$.

Under assumption b).

We may suppose that $N_j \leq \exp(C t^{\zeta} j^{\zeta(\alpha+\beta)})$.

We set $\mu = t^{-2\gamma}$, where $\zeta(1+2\gamma(\frac{\alpha+\beta}{\alpha-1})) = 2(1-\gamma)$, then we choose α large enough for $\gamma \geq \gamma(\zeta) - \frac{\eta}{2}$ and $\beta > 1$ to hold, where $\gamma(\zeta)$ is the solution of the above equation when $\alpha = +\infty$ (Namely : $2(1-\gamma(\zeta)) = k$). So :

$$A = O_{\eta,F}(1) \exp(O_{\eta,F}(1)\, t^{k+\eta}) \exp(-2t^2)$$

and

$$B = o_{\eta,F}(1) \exp(-2t^2)$$

whenever $t^2 \geq \frac{2}{\beta} + 5 + C t^{\zeta} 2^{2\beta+2}$.

So theorem 3.3. is proved in case 1°).

<u>Proof of theorem 3.3. in case 2°).</u>

We set $\varphi = \frac{t}{\sigma}$ and choose $m = \varphi^q$, $a = 2\varphi^{-q}$, $\mu = \varphi^{-q}$ and $\tau_j = \frac{\sigma}{\sqrt{n}} j^{-(\alpha+\beta)}$.

The variable A is this time controlled with the help of 3.1.2°), so now *the problem is to replace* σ_N^2 *by* σ_P^2 .

In fact, let Θ_N be the $(x_1,\dots,x_N)$-measurable event :

$\Theta_N = \{\|\sigma_N^2 - \sigma_P^2\| > s\}$, where $\sigma_N^2(f) = P_N(f^2) - (P_N(f))^2$ for any f in F .

Each term of the following estimate is studied in the sequel :

$$\Pr(\|\tilde{P}_n - P_N\| > \varepsilon') \leq \Pr(\Theta_N) + A' + B'$$

where $A' = E(A\, 1\!\!1_{\Theta_N^c})$ and $B' = E(B\, 1\!\!1_{\Theta_N^c})$.

Bounding $Pr(\Theta_N)$ *is a problem of type 1°)*

For, setting $F^2 = \{f^2, f \in F\}$, we have :

$$\|\sigma_N^2 - \sigma_P^2\| \leq \|P_N - P\|_{F^2} + 2\,\|P_N - P\|$$

Since $N_1^{(2)}(.,F^2) \leq N_1^{(2)}(\frac{\cdot}{2},F)$ and F^2 fulfills (M), we may use the bounds in 3.3.1°), so, choosing $s = \frac{2\varphi}{\sqrt{2nm}}$, we get : $Pr(\Theta_N) \leq C_o \exp(-\frac{\varphi^2}{2})$

The evaluation of A' and B'.

$\|\sigma_N^2\| \leq \sigma^2 + s$ holds on Θ_N^c , thus applying 3.1.2°) gives :

$$A' \leq 2\, N_r \exp(5\,\varphi^{2-q}) \exp\left(\frac{-t^2}{2\,(\sigma^2 + (\frac{t+\varphi^{1-q/2}}{\sqrt{n}}))}\right) , \quad \text{whenever } \varphi^q \geq 2 .$$

Moreover :

$$B' \leq 2 \sum_{j \geq r+1} N_j^2 \exp\left(-\frac{n\varepsilon'^2}{4\,\sigma^2}(j-1)^{2\beta}\right)$$

Now the proofs are completed as in case 1°), choosing this time $q=2$ and $\alpha = \text{Max}\,(2, 1 + \frac{4d}{3\eta})$ under assumption a) and $q = (2-\zeta)(\frac{\zeta}{2} + \zeta(\frac{\alpha+\beta}{\alpha-1}) + 1)^{-1}$ with α large enough for $1 - \frac{\zeta}{2} \leq p + \eta$ and $\beta > 1$ to hold.

To prove proposition 3.5 we choose $\alpha = 2$, so

$$A' \leq 2\, C\, \varepsilon_o^{-2d}\, e^5\, \sigma^{-2d}\, \varphi^{2d}\, (2+\varphi^2)^{6d} \exp\left(-\frac{t^2}{2\,(\sigma^2 + \frac{(3+t)}{\sqrt{n}})}\right),$$

$$B' \leq 2\, C^2\, \varepsilon_o^{-4d}\, \sigma^{-4d}\, \varphi^{4d}\, (2+\varphi^2)^{12d} \exp\left(-\frac{(\varphi^2-8)^2}{4}\right)$$

whenever $\varphi^2 \geq 8 + 12d$.

Besides, using 3.8.3., we get :

$$Pr(\Theta_N) \leq 2\, K_1 \left(\frac{\varepsilon_o}{2}\right)^{-4d} \left(\frac{\varphi}{2}\right)^{4d} \left(2 + \frac{\varphi^2}{2}\right)^{12d} \exp\left(-\frac{\varphi^2}{2}\right)$$

whenever $\frac{\varphi^2}{4} \geq 7+12d$, which completes the proof of proposition 3.5 via Lemma 3.2.

Proof of proposition 3.7.

We assume that $u=0$ and $v=1$. Inequality 3.3.2°)a) may be written :

$$M_o \ : \ \Pr(\|\nu_n\| > t) \le K\,\sigma^{-\alpha_1}\left(\frac{t^2}{\sigma^2}\right)^{\alpha_2} \exp\left(-\frac{t^2}{2\left(\sigma^2 + \frac{(3+t)}{\sqrt{n}}\right)}\right)$$

whenever $\frac{t^2}{\sigma^2} \ge 5$.

Defining the following sequences by induction :

$$a_{j+1} = 5 \ 4^{4d+1} \ a_j \left(2 + \frac{2\, b_j}{\sigma^2} + \frac{1}{n\sigma^2}\right)$$

$$b_{j+1} = \frac{2}{\sqrt{n}}\left(\frac{1}{\sqrt{n}} + \sqrt{b_j}\right)$$

with $a_o = 1$ and $b_o = \frac{3}{\sqrt{n}}$, we call M_j the following inequality :

$$\Pr(\|\nu_n\| > t) \le Ka_j\,\sigma^{-\alpha_1}\left(\frac{t^2}{\sigma^2}\right)^{\alpha_2} \exp\left(-\frac{t^2}{2\left(\sigma^2+b_j + \frac{t}{\sqrt{n}}\right)}\right) \text{ whenever } \frac{t^2}{\sigma^2} \ge 5 \quad .$$

The, assuming that M_j holds, it is possible to deduce M_{j+1} from M_j by the same way as 3.3.2°)a) from 3.3.1°)a) (technical details are given in [42]) .

Then inequalities (M_j) hold by induction. Using inequality M_J , where $J = 1 + [\frac{LLn}{L2}]$ and a few calculations yield proposition 3.7. .

4. EXPONENTIAL BOUNDS FOR THE BROWNIAN BRIDGE

We assume that $P(F^2) < \infty$. We want to show that the bounds in the preceding section still hold for the brownian bridge.

4.1. THEOREM

If $e_F^{(2)}(F) < 2$, then there exists some version G_P of a brownian bridge relating to P whose trajectories are uniformly continuous and bounded on (F,ρ_P) . Moreover, setting $\zeta = e_F^{(2)}(F)$, if $\|\sigma_P^2\|_F \le \sigma^2 \le P(F^2)$, an upper bound for $\Pr(\|G_P\|_F > t)$, is, for any positive t and η , given by :

$$0_{\eta,F}(1) \exp(0_{\eta,F})\, \sigma^{-\zeta-2\eta} \left(P(F^2))^{\zeta/2+\eta}(\tfrac{t}{\sigma})^{2p-\zeta+\eta} + (\tfrac{t}{\sigma})^{2p+\eta}\right) \exp\left(-\frac{t^2}{2\sigma^2}\right) \quad \textit{4.1.1.}$$

or, if more precisely $d_F^{(2)}(F) = 2d < \infty$, by :

$$0_{\eta,F}(1)\ (P(F^2))^{2d+\eta}\sigma^{-4d-2\eta}(1+\frac{t^2}{\sigma^2})^{2d+\eta}\ \exp\ (-\frac{t^2}{2\sigma^2}) \qquad \textit{4.1.2.}$$

where p is defined in the statement of theorem 3.3.

Comments.

In the framework of theorem 4.1. the existence of a regular version of a brownian bridge is an easy consequence of the proof of 4.1.1., but is of course a well known result (see [18]) . Moreover the bounds in 4.1. are in this case sharper than the more general Fernique-Landau-Shepp inequality (see [25]) that can be written :

$$\Pr\ (\ \|G_P\|_F > t) \leq C(\alpha)\ \exp\ (-\frac{t^2}{2\sigma^2}) \quad , \quad \text{for any}\quad \alpha > \|\sigma_P\|_F\ .$$

Proof of theorem 4.1.

If F is countable :

The calculations are similar to those of the proof of theorem 3.3. but here of course a sequence of nets in (F,F) *relating to P* is directly given . Moreover the following single inequality is used instead of Lemma 3.1. :

4.2. LEMMA :

Let V be a real and centered gaussian random variable with variance v^2 , then :

$$\Pr\ (|V| > s) \leq 2\ \exp\ (-\frac{s^2}{2v^2})\ \text{for any positive s.}$$

The choice of parameters being the same as in the proof of 3.3.2°) (except $\tau_j^2 = \frac{\sigma^2}{P(F^2)}\ j^{-2(\alpha+\beta)}$), 4.1.1. and 4.1.2. are proved. Since 4.1.1. is also an oscillation control, the almost sure regularity of G_P follows from Borel-Cantelli.

The general case.

Since (F,ρ_P) is separable , the familiar extension principle may be used to construct a regular version of brownian bridge on F from a regular version defined on a countable dense subset of F. Inequalities 4.1.1. and 4.1.2. still hold for this version.

Comment.

The optimality of bound 4.1.2. is discussed in the appendix. The degrees of the polynomial factors are different in 3.3.1.2°)a) and in 4.1.2. ; the reason is that bound 3.1.3°) is less efficient than bound 4.2. .

5. WEAK INVARIANCE PRINCIPLES WITH SPEEDS OF CONVERGENCE

We assume from now that $P(F^{2+\delta}) < \infty$ for some δ in $]0,1]$.

Using the results in sections 3 and 4, we can evaluate the oscillations of the empirical brownian bridge and of a regular version of the brownian bridge over F , so we can control the approximations of these processes by some E_k-valued processes (where E_k is a vector space with finite dimension k). The Prokhorov distance between the distributions of these two processes is estimated via an inequality from Dehling [15] allowing reasonable variations of k with n .

Oscillations of the empirical brownian bridge over F .

The oscillations of ν_n over F are controlled with the help of a truncation from 3.3.2°)a) (the proof in this case is straightforward) on the one hand and of a slight modification in the proof of 3.3.2°)b) (truncating twice) on the other hand. We shall not give any proof of the following theorem (the reader will find it in [42]) .

5.1. THEOREM.

We set $\mu_\delta = P(F^{2+\delta})$. If we assume that $\|\sigma_P^2\|_F \leq \sigma^2 \leq P(F^2)$ with $\sqrt{n}\sigma \geq 1$, then an upper bound for $Pr\,(\|\nu_n\|_F > t)$ is, for any positive t such that $t^2/\sigma^2 \geq 1$, given by :

a) If $d_F^{(2)}(F) = 2d < \infty$,

$$O_F(1)\, n^{7d} \left(\frac{t}{\sigma}\right)^{8d} \exp\left(-\frac{t^2}{8\sigma^2\,(2+O(\frac{LLn\sigma}{t}))}\right) + 128\, n^{-\delta/2}\sigma^{-2\delta}t^{-2+\delta}$$

whenever the following condition holds :

$$n^{\delta/2}\sigma^{2+2\delta}t^{-\delta} \geq 64\ \mu_\delta \qquad \underline{5.1.1.}$$

b) If $e_F^{(2)}(F) = \zeta < 2$

$$0_{\eta,F}(1)\ \exp\ (0_{\eta,F}(1)\ (\tfrac{t}{\sigma})^{2p+\eta}\ (1+ P(F^2))^{\frac{\zeta}{2}+\eta}\sigma^{-\zeta-2\eta}(\tfrac{t}{\sigma})^{-\zeta})\ \exp\ (-\ \frac{t^2}{24\sigma^2})$$

$$+\ 0_{\eta,F}(1)\ \exp\ (-\ \tfrac{1}{16}\ (\tfrac{t}{\sigma})^{2-2p+\eta}) + 0_\eta(1)\ (1+\mu_\delta)^2\ (n(\tfrac{t}{\sigma})^{2-2p-\eta})^{-\delta/2}\ (\sigma^{-2-\delta}+\sigma^{-\delta})$$

$$+\ 0(1)\ \mu_\delta\ n^{-\delta/2}\ \sigma^{-2\delta}t^{-2+\delta} \qquad \text{for any positive } \eta\ ;$$

(p is defined in the statement of 3.3.) whenever 5.1.1. and the following hold :

$$n^{\delta/4}\sigma^{2+\delta} \geq 512\ \mu_\delta\ . \qquad \underline{5.1.2.}$$

Remark.

Note that Yukich in [54] also used Kolčinskii-Pollard entropy conditions to prove analogous results to theorems 3.3. and 5.1., but our estimates are sharper because of the use of randomization from a large sample as described in section 3 .

Speed of convergence in the central limit theorem in finite dimension.

We recall below a result that is due to Dehling [15] (the first result in the same direction is due to Yurinskii [53]).

5.3. THEOREM.

Let $(X_i)_{1\leq i\leq n}$ be a sample of centered $\mathbb{R}^k$-valued random variables.

We write F_n for the distribution of the normalized sum of these variables and G for the centered gaussian distribution whose covariance is that of X_1 . Let $\|.\|_2$ be an euclidian pseudo-norm on $\mathbb{R}^k$ and π_2 be the Prokhorov distance that is associated to $\|.\|_2$. If $E\ (\|X_1\|_2^{2+\delta}) = \mu < \infty$, then :

$$\pi_2\ (F_n,G) \leq K\ n^{-\frac{\delta}{8}}\ k^{1/4}\ \mu^{1/4}\ (1\ +\ |L(n^{-\delta/2}\ k^{-1}\ \mu)|^{1/2})$$

where K is a universal constant.

Weak invariance principles for the empirical brownian bridge.

In order to build some regular versions of brownian bridges with given projection on a finite dimensional vector space (or further in section 6 on a countable product of such spaces), we need two lemmas.

5.3. LEMMA (Berkes, Philipp [6]) .

Let R_1, R_2, R_3 be Polish spaces, Q_1 and Q_2 be some distributions respectively defined on $R_1 x R_2$ and $R_2 x R_3$ with common marginal on R_2. Then there exists a distributions Q on $R_1 x R_2 x R_3$ whose marginals on $R_1 x R_2$ and $R_2 x R_3$ are respectively Q_1 and Q_2.

Remember that $l_T^\infty(F)$ is generally not separable. The following lemma is fundamental to avoid this difficulty (see [23]) . The space Ω to be mentionned below is defined in Section 2.

5.4. LEMMA (Skorohod [48])

Let R_1, R_2 be Polish spaces and Q be some distribution on $R_1 x R_2$ with marginal q on R_2. If V is a random variable from Ω to R_2 whose distribution is q, then there exists a random variable Y from Ω to R_1 such that the distribution of (Y,V) is Q.

Concerning our problems of construction the point in the sequel is that the distribution on $l_T^\infty(F)$ of a regular version of a brownian bridge is concentrated on a separable space.

Now we can state some weak invariance principles for the empirical brownian bridge with speeds of convergence.

5.5. NOTATIONS.

From now γ and β are positive functions that are respectively defined on $[0,1] \times \mathbb{R}_+$ and $[0,2]$ by :

$$\gamma(x,y) = \frac{x}{8 + 2y\,(4+x)} \qquad \text{and} \qquad \beta(z) = \frac{2(1-p(z))}{z(2-2p(z)+z)}$$

where, as in the statement of theorem 3.3. , $p(z) = \frac{2z(4-z)}{4 + z(4-z)}$

5.6. THEOREM

Under each of the following assumptions there exists some continuous version on (F,ρ_P) of a brownian bridge relating to P, $G_P^{(n)}$ such that $\Pr(\|\nu_n - G_P^{(n)}\|_F \geq \alpha_n) \leq \beta_n$ where (α_n) and (β_n) are defined hereunder (we recall that $F \in L^{2+\delta}(P)$ and that $e_F^{(2)}$ and $d_F^{(2)}$ are defined in Section 2) :

a) If $d_F^{(2)}(F) = 2d < \infty$

$$\alpha_n = \beta_n = O(n^{-\tau})$$

for any $\tau < \gamma(\delta,d)$

a') If $N_F^{(2)}(\varepsilon,F) \leq C\,\varepsilon^{-2d}(L\varepsilon^{-1})^d$ for any ε in $]0,1[$.

$$\alpha_n = \beta_n = O((Ln)^{(1/2)+d}\, n^{-\gamma(\delta,d)})$$

b) If $e_F^{(2)}(F) = \zeta < 2$

$$\alpha_n = O((Ln)^{-\tau}) \text{ and } \beta_n = O((Ln)^{-s})$$

for any $\tau < \beta(\zeta)$ and any positive s .

Proof of theorem 5.6.

Let σ be an oscillation rate (depending on n) and Π_σ be a projection of F on a σ-net $F(\sigma)$ relating to P .

We approximate ν_n uniformly over F by $\nu_n \circ \Pi_\sigma$.

Setting $G_\sigma = \{f-g\ ,\ \rho_P(f,g) \leq \sigma\}$, we may apply theorem 5.1. to G_σ (changing d into 2d if necessary), hence the quantity $\|\nu_n - \nu_n \circ \Pi_\sigma\|_F \leq \|\nu_n\|_{G_\sigma}$ can be evaluated. Besides, let $F_{n,\sigma}$ be the distribution of $\nu_{n|F(\sigma)}$ on the k-dimensional vector space $l^\infty(F(\sigma))$ and G_σ be the corresponding gaussian distribution.

Writing π_∞ for the Prokhorov distance associated to $\|\cdot\|_{F(\sigma)}$ and applying Strassen's theorem [49] , there exists a probability space $(\Omega',A',\Pr')$ and two random variables $\nu_n(\sigma)$ and $G(\sigma)$ with respective distributions $F_{n,\sigma}$ and G_σ on $l^\infty(F(\sigma))$ such that :

$$\Pr'(\|\nu_n(\sigma) - G(\sigma)\|_{F(\sigma)} \geq B) \leq B$$

where $B = \pi_\infty(F_{n,\sigma}, G_\sigma)$.

So, using lemma 5.3. , we may ensure the existence of some continuous version of a brownian bridge G_P relating to P such that $G(\sigma) = G_{P|F(\sigma)}$ and then, applying lemma 5.4. with $V : \omega \to \nu_{n|F(\sigma)}$, we may assume that G_P is constructed on Ω with $Pr(\|\nu_{n|F(\sigma)} - G_{P|F(\sigma)}\|_{F(\sigma)} \geq B) \leq B$.

Hence, noticing that $\|\nu_{n|F(\sigma)} - G_{P|F(\sigma)}\|_{F(\sigma)} = \|(\nu_n - G_P)\circ\Pi_\sigma\|_F$, we get :

$$Pr\ (\|\nu_n - G_P\|_F > 2t+B) \leq A+B+C$$

where $A = Pr\ (\|\nu_n\|_{G_\sigma} > t)$ and $C = Pr\ (\|G_P\|_{G_\sigma} > t)$.

Theorem 5.2. is used to control B (with $\|\cdot\|_{F(\sigma)} \leq \|\cdot\|_2 \leq \sqrt{k}\,\|\cdot\|_{F(\sigma)}$) noticing that $k \leq N_F^{(2)}\ (\frac{\sigma}{2},F)$ according to remark 2.6. .

Moreover C is evaluated with the help of theorem 4.1., so the calculations are completed via an appropriate choice of t and σ .

6. STRONG INVARIANCE PRINCIPLES WITH SPEEDS OF CONVERGENCE.

The method to deduce strong approximations from the preceding weak invariance principles is the one used in [43] to prove theorem 2 : the weak estimates are used locally, giving strong approximations with the help of maximal inequalities and via Borel-Cantelli lemma.

Maximal inequalities.

As was noticed in [23], the proofs of the following inequalities may be deduced from the one given in [10] and in [32] .

Notation.

We set $X_j = \delta_{x_j} - P$ for any integer j .

6.1. LEMMA *(Ottaviani's inequality).*

We set $S_k = \sum_{j=1}^{k} X_j$. Then, for any positive α , the following inequality holds:

$$(1-c)\ \Pr\left(\max_{k\leq n} \|S_k\|_F > 2\alpha\right) \leq \Pr\left(\|S_n\|_F > \alpha\right)$$

where $c = \max_{k\leq n} \Pr\left(\|S_k\|_F > \alpha\right)$.

More precisely, for symmetrical variables, the following sharper inequality is available :

6.2. LEMMA *(Paul Lévy's inequality).*

Let $(Y_i)_{1\leq n}$ be independent and identically distributed B-valued random variables where $(B, \|.\|)$ is a normed vector space. If we assume that Y_1 is symmetrical then : $\Pr(\max_{k\leq n} \|S_k\| > \alpha) \leq 2 \Pr(\|S_n\| > \alpha)$ holds for any positive α , where $S_k = \sum_{j=1}^{k} Y_j$.

Strong approximations for the empirical brownian bridge.

6.3. THEOREM.

Under each of the following assumptions some sequence $(Y_j)_{j\geq 1}$ of independent versions of brownian bridges relating to P that are continuous on (F,ρ_P), may be defined on Ω such that :

a) if $d_F^{2)}(F) = 2d < \infty$,

$$\frac{1}{\sqrt{n}} \left\| \sum_{j=1}^{n} (X_j-Y_j) \right\|_F = O(n^{-\alpha}) \qquad \text{a.s.}$$

for any $\alpha < \dfrac{\gamma(\delta,d)}{2(1+\gamma(\delta,d))}$;

a') if, more precisely , $N_F^{(2)}(\varepsilon,F) \leq C\varepsilon^{-2d}(1+L\varepsilon^{-1})^d$ for any ε in $]0,1[$,

$$\frac{1}{\sqrt{n}} \left\| \sum_{j=1}^{n} (X_j-Y_j) \right\|_F = O(n^{-\gamma(\delta,d)/(2(1+\gamma(\delta,d)))}\ ((Ln)^{(1/2)+d} + (Ln)^{(5/4)+(d/2)}))\ \text{a.s.} ;$$

b) if $e_F^{(2)}(F) = \zeta < 2$,

$$\frac{1}{\sqrt{n}} \left\| \sum_{j=1}^{n} (X_j-Y_j) \right\|_F = O(Ln^{-(\beta/2)}) \qquad \text{a.s.}$$

for any $\beta < \beta(\zeta)$.

Where $\gamma(.,.)$ and $\beta(.)$ are defined in 5.5.

For a proof of 6.3., see [42].

Comments.

When passing from weak invariance principles to strong ones, the speeds of convergence are transformed as follows within our framework :

(i) $n^{-\gamma} \to n^{-\gamma/(2(1+\gamma))}$

in case a).

(ii) $Ln^{-\beta} \to Ln^{-\beta/2}$

in case b) .

Transformation (ii) appears in theorem 6.1. (under 6.3 .) from [23], it is not the case for transformation (i) in the same theorem (under 6.4.).

On the contrary transformation (i) is present in finite dimensional principles and appears to be optimal in that case : more precisely, the rate of weak convergence towards the gaussian distribution for 3-integrable variables is ranging about $n^{-1/2}$ when the rate of strong convergence is ranging about $n^{-1/6}$ (see [39] for the upper bound and [9] for the lower bound), in the real case .

Application to V.Č.-classes.

Applying theorem 6.3. with $\delta = 1$ in the case where F is a V.Č.-class with real density d, we get a speed of convergence towards the brownian bridge that is $O(n^{-\alpha})$ for any $\alpha < \frac{1}{18+20d}$. This improves on 1.3.3. but is less sharp that 1.2.4. in the classical case of quadrants in $\mathbb{R}^d$.

6.4. INVARIANCE PRINCIPLES IN C(S).

Following an idea from Dudley in [21] (sec.11), the study of the general empirical processes theoretically allows one to deduce some results about random walks in general Banach spaces. As an application of this principle let us consider a compact metric space (S,κ) and the space $C(S)$ of real continuous functions on S, equipped with the uniform norm $\|\cdot\|_\infty$. Let X be the space of Lipschitz-functions on S equipped with the Lipschitz-norm :

$$\|\cdot\|_L : x \to \|x\|_\infty + \sup_{t\neq s} \frac{|x(t)-x(s)|}{\kappa(s,t)}$$

We write $N(\varepsilon,S,\kappa)$ for the maximal cardinality of a subset R of S such that $\kappa(s,t) > \varepsilon$ for any $s\neq t$ in R.

We may apply our results through the following choices :

$$F = \{\delta_s, s\in S\} \text{ and } F = \|\cdot\|_L .$$

Then $(X, \|\cdot\|_\infty)$ is a Suslin space (but is not Polish in general), so F fulfills (M). Moreover, for any distribution Q in $P_F^{(2)}(X)$ we have :

$$Q((\delta_s-\delta_t)^2) \leq \kappa^2(s,t)\, Q(F^2) ,$$

so $N_F^{(2)}(.,F) \leq N(.,S,\kappa)$. Besides $\|\cdot\|_\infty = \|\cdot\|_F$.

Therefore, considering a sequence $(X_j)_{j\geq 1}$ of independent and identically distributed $C(S)$-valued random variables such that :

$$|X_1(s)-X_1(t)| \leq M\, \kappa(s,t) \quad , \text{ for any } s,t \text{ in } S .$$

with $E(M^{2+\delta}) < \infty$ and $E(X_1^{2+\delta}(t_0)) < \infty$ for one t_0 in S, we can apply some 5.5. or 6.3. theorem to get speeds of convergence towards the gaussian distribution, whose structure depends on $N(.,S,\kappa)$ (the central limit theorem for such uniformly Lipschitzian processes as above is due to Jain and Marcus in [31]).

APPENDIX

1. PROOF OF LEMMA 3.1.

First let us recall Hoeffding's lemma (see [29]).

Hoeffding's lemma.

Let S be a centered and $[u,v]$-valued random variable, then :

$$E(\exp(tS)) \leq \exp\left(\frac{t^2(v-u)^2}{8}\right) \quad , \text{ for any } t \text{ in } \mathbb{R} .$$

We may assume that w is chosen as follows :

. drawing - with uniform distribution - a partition $J = (J_i)_{1\leq i\leq n}$ such that $|J_i| = m$ for each i in $[1,n]$.

. then, drawing an index $w(i)$ independently in each J_i - with uniform distribution - . The following evaluations are conditional on $\mathcal{J}$ but the last bound will not depend on $\underset{\sim}{\mathcal{J}}$, giving 3.1. .

We set $Z = \frac{S_n}{n} - \frac{S_N}{N}$ and we write Λ for the logarithm of the conditional Laplace transform of Z .

Then setting $\bar{\xi}_i = \frac{1}{m} \sum_{j \in J_i} \xi_j$, we have , for any s in $\mathbb{R}$:

$$\Lambda(s) = \sum_{i=1}^{n} \operatorname{Log} \left(\frac{1}{m} \sum_{j \in J_i} \exp\left(\frac{s}{n}(\xi_j - \bar{\xi}_i)\right)\right)$$

then, since the logarithm is a concave function :

$$\Lambda(s) \leq n \operatorname{Log} \left(\frac{1}{N} \sum_{j=1}^{N} \exp\left(\frac{s}{n}\left(\xi_j - \frac{S_N}{N}\right)\right)\right) = n\, \Lambda_N\left(\frac{s}{n}\right)$$

where, writing Q_N for the uniform distribution on $\{\xi_1,\dots,\xi_N\}$, Λ_N stands for the logarithm of the Laplace transform under Q_N of $x \to x - E_{Q_N}(x)$. Therefore the Cramer-Chernoff transform of Z is larger than that of $\frac{S'_n}{n} - E_{Q_N^{\otimes n}}\left(\frac{S'_n}{n}\right)$ under $Q_N^{\otimes n}$ where S'_n stands for the sum of n i.i.d. random variables with common distribution Q_N .

Then, Hoeffding [29] and Bernstein [5] inequalities yield 3.1.1°) and 3.1.2°) . In order to prove 3.1.3°) we may assume that $S_N = 0$ (otherwise changing ξ_j into $\xi_j - \frac{S_N}{N}$) .

Then, applying Hoeffding's lemma to the conditionally centered random variables $\xi_{w(i)} - \bar{\xi}_i$ and setting $u_i = \min_{j \in J_i} \xi_j$ and $v_i = \max_{j \in J_i} \xi_j$, we get :

$$l_i(t) = \operatorname{Log} E^{\mathcal{J}}\left(\exp\left(t(\xi_{x(i)} - \bar{\xi}_i)\right)\right) \leq \frac{(v_i - u_i)^2}{8}\, t^2 \text{ , for any } t \text{ in } \mathbb{R}\,.$$

Hence

$$\Lambda(s) = \sum_{i=1}^{n} l_i\left(\frac{s}{n}\right) \leq \frac{s^2}{8n^2} \sum_{i=1}^{n} (v_i - u_i)^2 \leq \frac{s^2}{4n^2} \sum_{i=1}^{n} \sum_{j \in J_i} \xi_j^2$$

and therefore :

$$\Lambda(s) \leq \frac{s^2}{4n}\, m\, \sigma_N^2$$

yielding 3.1.3°) via Markov's inequality.

2. THE DISTRIBUTION OF THE SUPREMUM OF A d-DIMENSIONAL PARAMETER BROWNIAN BRIDGE.

We use Goodman's work in [28] to give a lower bound of the probability for the supremum of a brownian bridge to cross a barrier.

Notations.

We set $I = [0,1]$ and write for any integer d, 1_d for the element $(1,\dots,1)$ of $\mathbb{R}^d$. Moreover, for any s in I^d, we set $p(s) = s_1 \dots s_d$.

A.1. Theorem.

Let d be an integer and W_d be some standard d-dimensional parameter Wiener process, then, on the one hand :

(i) $$\Pr\left(\sup_{s\in I^d} W_d(s) \leq t \mid W_d(1_d) = at\right) \leq h_d(a,t)$$

for almost any real number a (in Lebesgue sense) and any positive t, where

$$h_d(a,t) = 1 + \exp(2t^2(a-1)) \sum_{i=0}^{d-1} (-1)^{i+1} \frac{(2t^2(a-1))^i}{i!} I_{]-\infty,1]}(a)$$

and on the other hand :

(ii) $$\Pr\left(\sup_{s\in I^d} W_d(s) - p(s)\, W_d(1_d) > t\right) \leq \sum_{i=0}^{d-1} \frac{(2t^2)^i}{i!} \cdot \exp(-2t^2) \quad .$$

Proof of theorem A.1.

If $d=2$ the whole proof is contained in [28] .

Otherwise, proceeding exactly as in [28] yields the following inequality :

$$H_d(a,t) = \Pr\left(\sup_{s\in I^d} W_d(s) \leq t \mid W_d(1_d)=at\right) \leq \int_{a-1}^{0} (1-\exp(2t^2 r))\, dF_{t,d-1}(a,dr) \qquad \text{A.2.}$$

where $F_{t,d-1}(a,r) = \Pr(W_{d-1}(s) - rtp(s) \leq t,\ \forall s\in I^{d-1} \mid W_{d-1}(1_{d-1})=at)$.

We want to proceed by induction. It is enough to notice that :

A.3. Lemma.

$$\Pr(W_k(s)+\alpha p(s) \leq t,\ \forall s\in I^k \mid W_k(1_k)+\alpha = \beta) = \Pr(W_k(s) \leq t,\ \forall s\in I^k \mid W_k(1_k) = \beta)$$

for any integer k, any positive α and almost every β in $\mathbb{R}$ (in Lebesgue sense).

Proof of A.3.

Since W_k is a regular gaussian process, it is enough to show that the expectation and covariance functions of the processes $W_k(.)+p(.)$ and $W_k(.)$ are the same conditionally to respectively $W_k(1_k)+\alpha$ and $W_k(1_k)$.

Since W_k is gaussian, $E(W_k(s)\,|W_k(1_k)=\gamma)$ and $E(W_k(s)W_k(s')|W_k(1_k)=\gamma)$ are respectively linear and quadratic functions of γ, then the knowledge of $E(W_k^l(1_k)\; W_k^m(s)\; W_k^n(s'))$ with $l+m+n \le 4$, yields :

$$E(W_k(s)|W_k(1_k)=\gamma) = p(s)\gamma \ ,$$

$$E(W_k(s)W_k(s')|W_k(1_k)=\gamma) = p(s)p(s')\gamma^2 + (s\wedge s') - p(s)p(s')$$

where $s\wedge s' = (s_1\wedge s_1')\dots(s_k\wedge s_k')$, A.3. .

Let us return to the proof of A.1.(i) .

Using lemma A.3., we get :

$$F_{t,d-1}(a,r) = H_{d-1}(a-r,t)$$

so :

$$F_{t,d-1}(a,r) \le h_{d-1}(a-r,t)$$

Then, integrating by parts, inequality A.2. becomes :

$$H_d(a,t) \le F_{t,d-1}(a,0) - |\exp(2t^2r)F_{t,d-1}(a,r)\,|_{a-1}^{0} + 2t^2\int_{a-1}^{0} \exp(2t^2r)F_{t,d-1}(a,r)\,dr$$

hence

$$H_d(a,t) \le 2t^2\int_{a-1}^{0} \exp(2t^2r)\; h_{d-1}(a-r,t)\,dr \quad .$$

But an easy calculation yields :

$$2t^2\int_{a-1}^{0} \exp(2t^2r)\; h_{d-1}(a-r,t)\,dr = h_d(a,t) \quad .$$

Therefore A.1. (i) is proved by induction (it is shown in [33] p. 284 that inequality A.1.(i) holds when $d=1$) .

In order to proof (ii), we notice, following [7], p.84, that :

$$\Pr(W_d\in\cdot|\; 0 \le W_d(1_d) \le \varepsilon\,)$$

converges weakly in $C(I^d)$ towards the distribution of the brownian bridge :

$W_d - p(.)W_d(1_d)$ whenever ε converges to 0 .

So, inequality (i) gives :

$$\Pr\,(\sup_{s\in I^d} W_d(s) - p(s)W_d(1_d) > t) \geq 1 - h_d(0,t)$$

therefore (ii) is proved.

Comment.

Theorem A.1. was proved by ourself (see [40] and [41]) but also by E. Cabaña in [11]* .

In another connection, inequality A.1. (ii) ensures that some polynomial factor $t^{2h(d)}$ with $h(d) \geq d-1$ cannot be removed in bounds 3.3.1°)a) and 4.1.2. .

3. EXPLICITING AN EXPONENTIAL BOUND.

The calculations yielding 3.3.1°)a) are slightly modified here, where the entropy condition a) is replaced with a more explicit one.

A.4. Theorem.

If we assume that F is $[0,1]$-valued and that

a') $N_1^{(2)}(\varepsilon,F) \leq K^{1+1/\mathrm{Log}(\varepsilon^{-2})}\,(1+\mathrm{Log}(\varepsilon^{-2}))^d\,\varepsilon^{-2d}$ for any ε in $]0,1[$

then, an upper bound for $\Pr\,(\|\nu_n\|_F > t)$ is, for any t in $[1,+\infty[$, given by :

$$4H(t)\exp(13)\exp(-2t^2) + 4H^2(t)\exp(-(t^2-5)(Lt)^2)$$

where

$$H(t) = K^{6/5}\exp(16d)\,(1+Lt^2)^{5d}\,t^{6d} \quad .$$

Proof of A.4.

In the proof of 3.3.1°)a) we choose $\alpha = \dfrac{Lt^2}{LLt^2} + 1$, then

$$A \leq 2H(t)\exp(13)\exp(-2t^2) \qquad P^{\otimes N}\text{-a.s.}$$

$$B \leq 2H^2(t)\exp(-(t^2-5)(Lt)^2) \qquad P^{\otimes N}\text{-a.s.}$$

whenever $t^2 \geq 6+4d$, yielding A.4. via lemma 3.2. .

---

* Thanks to M. Wchebor and J. Leon for communicating this reference to us.

Comment.

Assumption a') is typically fulfilled whenever F is a V.Č.-class. In that case d may be the real density of F (if it is "achieved") or the integer density of F (see the proof of 3.6.).

REFERENCES

[1] ALEXANDER, K. (1982). *Ph. D. Dissertation, Mass. Inst. Tech.* (1982).

[2] ALEXANDER, K. *Probability inequalities for empirical processes and a law of iterated logarithm. Annals of Probability* (1984), vol.12, N°4, 1041-1067.

[3] ASSOUAD, P. *Densité et dimension, Ann. Inst. Fourier, Grenoble* 33, 3 (1983) 233-282.

[4] BAKHVALOV, N.S. *On approximate calculation of multiple integrals (in Russian). Vestnik Mosk. Ser. Mat. Mekh. Astron. Fiz. Khim.* (1959) 4, 3.18.

[5] BENNETT, G. *Probability inequalities for sums of independent random variables. J. Amer. Statist. Assoc.* (1962), 57, 33-45.

[6] BERKES, I., PHILIPP, W. *Approximation theorems for independent and weakly dependent random vectors. Ann. Probability* (1979), 7, 29-54.

[7] BILLINGSLEY, P. *Convergence of probability measures. Wiley, New York.*

[8] BORISOV, I.S. *Abstracts of the Colloquium on non parametric statistical inference, Budapest* (1980), 77-87.

[9] BREIMAN, L. *On the tail behavior of sums of independent random variables. Z. Warschein. Verw. Geb.* (1967), 9, 20-25.

[10] BREIMAN, L. *Probability. Reading Mass. Addison-Wesley* (1968).

[11] CABAÑA, E. *On the transition density of a multidimensional parameter Wiener process with one barrier. J. Appl. Prob.* (1984), 21, 197-200.

[12] CABAÑA, E., WSCHEBOR, M. *The two-parameter Brownian bridge. Pub. Univ. Simon Bolivar.*

[13] COHN, D.L. (1980). *Measure theory. Birkhaüser, Boston* (1980).

[14] CSÖRGO, M. , RÉVÉSZ, P. *A new method to prove Strassen type laws of invariance principle II. Z. Warschein. Verw. Geb.* (1975), 31, 261-269.

[15] DEHLING, H. *Limit theorems for sums of weakly dependent Banach space valued random variables. Z. Warschein. Verw. Geb.* (1983), 391-432.

[16] DEVROYE, L. *Bounds for the uniform deviations of empirical measures. J. of Multivar. Anal.* (1982), 12, 72-79.

[17] INCHI HU, *A uniform bound for the tail probability of Kolmogorov-Smirnov statistics. The Annals of Statistics* (1985), Vol.13, N°2, 821-826.

[18] DUDLEY, R.M. *The sizes of compact subsets of Hilbert space and continuity of Gaussian processes.* J. Functional Analysis (1967), 1, 290-330.

[19] DUDLEY, R.M. *Metric entropy of some classes of sets with differential boundaries.* J. Approximation Theory (1974), 10, 227-236.

[20] DUDLEY, R.M. *Central limit theorems for empirical measures.* Ann. Probability (1978), 6, p. 899-929 ; correction 7 (1979), 909-911.

[21] DUDLEY, R.M. *Saint Flour 1982* . Lecture Notes in Mathematics n°1097 .

[22] DUDLEY, R.M., DURST . *Empirical Processes, Vapnik-Červonenkis classes and Poisson processes.* Proba. and Math. Stat. (Wroclaw) 1, 109-115 (1981).

[23] DUDLEY, R.M., PHILIPP, W. *Invariance principles for sums of Banach spaces valued random elements and empirical processes.* Z. Warschein. Verw. Geb. (1983), 82, 509-552.

[24] DVORETZKY, A., KIEFER, J.C., WOLFOWITZ, J. *Asymptotic minimax character of the sample distribution function and of the classical multinomial estimator.* Ann. Math. Stat. (1956), 33, 642-669.

[25] FERNIQUE, X. *Régularité de processus gaussiens.* Invent. Math. (1971), 12, 304-320.

[26] GAENSSLER, P., STUTE, W. *Empirical processes : a survey of results for independent and identically distributed random variables.* Ann. Proba. 7, 193-243.

[27] GINE, M.E., ZINN, J. *On the central limit theorem for empirical processes.* Annals of Probability (1984), vol. 12, n° 4, 929-989.

[28] GOODMAN, V. *Distribution estimations for functionals of the two parameter Wiener process.* Annals of Probability (1976), vol. 4, n° 6, 977-982.

[29] HOEFFDING, W. *Probability inequalities for sums of bounded random variables.* J. Amer. Statist. Assoc. (1963), 58, p. 13-30.

[30] IBRAGIMOV, I.A., KHASMINSKII, R.Z. *On the non-parametric density estimates.* Zap. Naucha. Semin. LOMI 108, p. 73-81 (1981). In Russian.

[31] JAIN, N., MARCUS, M.B. *Central limit theorem for C(S)-valued random variables.* J. Functional Analysis (1975), 19, p. 216-231.

[32] KAHANE, J.P. *Some random series of functions.* Lexington, Mass. D.C. Heuth (1968).

[33] KARLIN, S. TAYLOR, H.M. *A first course in Stochastic Processes* (1971). Academic Press, New-York.

[34] KIEFER, J.C. *On large deviations of the empirical d.f. of vector chance variables and a law of iterated logarithm.* Pacific J. Math. (1961), 11, 649-660.

[35] KIEFER, J.C., WOLFOWITZ, J. *On the deviations of the empiric distribution function of vector chance variables.* Trans. Amer. Math. Soc. (1958), 87, p. 173-186.

[36] KOLMOGOROV, A.N., TIKHOMIROV, V.M. *ε-entropy and ε-capacity of sets in functional spaces*. Amer. Math. Soc. Transl. (Ser. 2) (1961), 17, 277-364.

[37] KOMLOS, J., MAJOR, P., TUSNADY, G. *An approximation of partial sums of independent RV's and the sample DF. I.* Z. Warschein. Verw. Geb. (1975), 32, p. 111-131.

[38] LE CAM, L. (1983). *A remark on empirical measures.* In A Festscheift for Erich L. Lehmann in Honor of his Sixty-Fifth Birthday (1983), 305-327 Wadsworth, Belmont, California.

[39] MAJOR, P. *The approximation of partial sums of independent RV's.* Z. Warschein. Verw. Geb. (1976), 35, 213-220.

[40] MASSART, P. *Vitesse de convergence dans le théorème de la limite centrale pour le processus empirique.* Thèse de 3e cycle n° 3545 de l'Université de Paris-Sud (1983).

[41] MASSART, P. *Vitesses de convergence dans le théorème central limite pour des processus empiriques.* Note aux C.R.A.S., t. 296 (20 juin 1983) Série I, 937-940.

[42] MASSART, P. *Rates of convergence in the central limit theorem for empirical processes* (April 1985), submitted to Ann. Inst. Henri Poincaré.

[43] PHILIPP, W. *Almost sure invariance principles for sums of B-valued random variables.* Lecture Notes in Mathematics 709, 171-193.

[44] POLLARD, D. *A central limit theorem for empirical processes.* J. Australian Math. Soc. Ser. A (1982), 33, 235-248.

[45] POLLARD, D. *Rates of strong uniform convergence* (1982). Preprint.

[46] SERFLING, R.J. *Probability inequalities for the sum in sampling without replacement.* Ann. Stat. (1974), vol. 2, n° 1, 39-48.

[47] SION, M. *On uniformization of sets in topological spaces.* Trans. Amer. Math. Soc. (1960), 96, 237-245.

[48] SKOROHOD, A.V. Theory Prob. Appl. (1976), 21, 628-632.

[49] STRASSEN, V. *The existence of probability measures with given marginals.* Ann. Math. Stat. (1965), 36, 423-439.

[50] TUSNADY, G. *A remark on the approximation of the sample DF in the multi-dimensional case.* Periodica Math. Hung. (1977), 8, 53-55.

[51] VAPNIK, V.N. ČERVONENKIS, A.Y. *On the uniform convergence of relative frequencies of events to their probabilities.* Theor. Prob. Appl. (1971), 16, 264-28 .

[52] YURINSKII, V.V. *A smoothing inequality for estimates of the Lévy-Prohorov distance.* Theory Prob. Appl. (1975), 20, 1-10.

[53] YURINSKII, V.V. *On the error of the gaussian approximation for convolutions.* Theor. Prob. Appl. (1977), 22, 236-247.

[54] YUKICH, J.E. *Uniform exponential bounds for the normalized empirical process* (1985) . Preprint.

MEAN SQUARE CONVERGENCE OF WEAK MARTINGALES

Mariola B. Schwarz
Institut für Mathematische Statistik
Universität Göttingen
Lotzestr. 13, D-3400 Göttingen

In [4] it was shown that the mean square convergence of vector-valued martingales in spaces of Rademacher type or cotype 2 is closely related to the following property of a Banach space valued martingale $f = (f_n)$: there exists a Gaussian Radon measure γ_f on $(B,\mathcal{B})$ ($\mathcal{B}$ the Borel σ-algebra of subsets of the Banach space B) such that

(*) $$\|x^* f\|_2^2 = \int_B |x^*(x)|^2 \gamma_f(dx)$$

for every $x^* \in B^*$.

We give a characterization of the class of martingales satisfying (*) for spaces not containing l_∞^n uniformly and having an unconditional basis (e_n) in means of convergence of the series $\sum_{n=1}^{\infty} \|S(e_n^* f)\|_2 e_n$, where (e_n^*) is the dual basis and S is the standard square function.

Furthermore we give necessary (resp. sufficient) conditions for the L_2-convergence of martingales in spaces of type 2 (resp. cotype 2).
These conditions characterize Banach spaces of Rademacher type or cotype 2 .

Throughout, B denotes a separable Banach space, B^* the dual space, $\mathcal{B}$ the Borel σ-algebra of subsets of B and $(\Omega,\mathfrak{F},P)$ a probability space.

A family $\{f_n \mid n \in N\}$ of B-valued random variables forms a <u>weak martingale</u> with respect to the filtration $\{\mathfrak{F}_n \mid n \in N\}$ if for every $n \in N$, f_n is $\mathfrak{F}_n$ measurable, Pettis integrable on $\mathfrak{F}_n$ and (Pettis-) $E(f_n \mid \mathfrak{F}_k) = f_k$ for $k \le n$ (see e.g. [3]).
A B-valued weak martingale $\{f_n \mid n \in N\}$ is uniformly bounded if there is a constant M such that the inequality $|x^* f_n| \le M \|x^*\|$ holds a.e. for every $x^* \in B^*$ and every $n \in N$.
Note that if $f = \{f_n \mid n \in N\}$ is a B-valued (weak) martingale then $x^* f = x^* f = \{x^* f_n \mid n \in N\}$ is a real martingale for each $x^* \in B^*$.
In the following $f = \{f_n \mid n \in N\}$ is a B-valued uniformly bounded weak martingale with difference sequence $\{d_n \mid n \in N\}$.

LEMMA 1. <u>Let $(e_n)_{n\in N}$ be an unconditional basis in B and $(e_n^*)_{n\in N}$ the dual basis. Assume that B does not contain l_∞^n uniformly.</u>
<u>Then the following conditions are equivalent</u>

a) <u>the series</u> $\Sigma\|S(e_n^*f)\|_2\, e_n$ <u>is convergent in</u> B ;

b) <u>there exists a Gaussian measure</u> γ_f <u>on</u> $(B,\mathcal{B})$ <u>such that for each</u> $x^* \in B^*$

$$\|x^*f\|_2^2 = \int_B |x^*(x)|^2\, \gamma_f(dx) .$$

<u>Proof</u>. $x^*f = (x^*f_n)$ is for each $x^* \in B^*$ a real martingale with increments x^*d_n. Thus the square function S of x^*f is given by

$$S(x^*f) = (\sum_k |x^*(d_k)|^2)^{1/2} .$$

Let the functional $T : B^* \times B^* \to R$ be given by

$$T(x^*,y^*) = E\sum_k x^*(d_k)\, y^*(d_k) , \quad x^*,y^* \in B^* .$$

T is well defined everywhere in the Cartesian product $B^* \times B^*$ since f is of weak second order. Furthermore, T is positive (i.e., $T(x^*,x^*) \geq 0$ for every $x^* \in B^*$) and symmetric (i.e., $T(x^*,y^*) = T(y^*,x^*)$ for all $x^*,y^* \in B^*$). Since f is uniformly bounded we have

$$\begin{aligned} T(x^*,y^*) &\leq (E\sum_k x^*(d_k)^2)^{1/2} (E\sum_k y^*(d_k)^2)^{1/2} \\ &= \|S(x^*f)\|_2\, \|S(y^*f)\|_2 \\ &= \|x^*f\|_2\, \|y^*f\|_2 \\ &\leq M^2\|x^*\|\, \|y^*\| , \end{aligned}$$

which means that T is bounded.

We can consider T as a map of B^* into B^{**}. Using a theorem of Banach (see e.g. [1]), which states that an element u of B^* belongs to the image of B in the natural embedding of B into B^{**}, if the functional $u(v)$, $v \in B^*$, is continuous in the B-topology on B^*, we show that TB^* is contained in B.

Assume that a) is satisfied. Since

$$\sum_n \|S(e_n^*f)\|_2\, e_n = \sum_n (E(\sum_k(e_n^*(d_k))^2)^{1/2}\, e_n = \sum_n (Te_n^*(e_n^*))^{1/2}\, e_n$$

the series $\sum_n (Te_n^*(e_n^*))^{1/2}\, e_n$ is convergent in B.

By theorem 2.1. in [2], T is a covariance operator of a Gaussian measure γ on B, i.e.,

$$Tx^*(x^*) = \int_B |x^*(x)|^2\, \gamma(dx) .$$

Since

$$\|x^*f\|_2^2 = \|S(x^*f)\|_2^2 = Tx^*(x^*)$$

the implication a) $\Rightarrow$ b) follows.

Conversely, if b) is satisfied, the series

$$\sum_n \|S(e_n^*f)\|_2\, e_n = \sum_n (Te_n^*(e_n^*))^{1/2}\, e_n$$

is convergent in B as T is a covariance operator of γ_f. □

We recall that a Banach space does not contain l_∞^n uniformly if it is of certain Rademacher cotype $r < \infty$. Banach spaces of Rademacher type 2 are of some Rademacher cotype $r < \infty$ (see, e.g., [5]).

Using Theorem 5.3. and 5.4. of [4], Lemma 1 and the above remark we get

THEOREM 1. Let $(e_n)_{n\in N}$ be an unconditional basis in B and $(e_n^*)_{n\in N}$ the dual basis. Then

i) B is of cotype 2 if and only if for every f the convergence of the series $\sum_n \|S(e_n^* f)\|_2 e_n$ in B is a sufficient condition for the L^2 convergence of f.

ii) B is of type 2 if and only if for every f the convergence of the series $\sum_n \|S(e_n^* f)\|_2 e_n$ in B is a necessary condition for the L^2 convergence of f. ▭

COROLLARY 1. In Hilbert space (and only in this space) the convergence of the series $\sum_n \|S(e_n^* f)\|_2 e_n$ is a necessary and sufficient condition for a martingale f to converge in the L^2-norm. ▭

Let $\mathcal{M}_\gamma$ be the family of all B-valued weak martingales $f = (f_n)$ for which there exists a Gaussian measure γ_f on $(B,\mathcal{B})$ such that

$$\|x^* f\|_2^2 = \int_B |x^*(x)|^2 \gamma_f(dx)$$

for each $x^* \in B^*$.

Using Theorem 2.2. in [2] and the techniques of the proof of Lemma 1 we get

THEOREM 2. Let $(e_n)_{n\in N}$ be an unconditional basis in B, $(e_n^*)_{n\in N}$ the dual basis and $p > 0$. Assume that B does not contain l_∞^n uniformly. Then there exist constants $c_1(p)$ and $c_2(p)$ such that for each $f \in \mathcal{M}_\gamma$

$$c_1(p) \|\sum_n \| S(e_n^* f) \|_2 e_n\|^p \le \int_B \|x\|^p \gamma(dx)$$

$$\le c_2(p) \|\sum_n \| S(e_n^* f)\|_2 e_n\|^p .$$

REFERENCES

[1] S. BANACH : Théorie des opérations linéaires, Warszawa 1932.

[2] S.A. CHOBANYAN AND V.I. TARIELADZE : Gaussian characterizations of certain Banach spaces. J. Mult. Anal. 7, 1977.

[3] K. MUSIAL : Martingales of Pettis integrable functions. Lecture Notes in Math. 794, 1979.

[4] NGUYEN DUY TIEN : On Kolmogorov's three series theorem and mean square convergence of martingales in Banach spaces. Theor. Prob. Appl. 24(2), 1979.

[5] W.A. WOYCZYNSKI : Geometry and martingales in Banach spaces. Advances in Prob., Dekker, 1978.

METRIC ENTROPY AND THE CENTRAL LIMIT THEOREM IN BANACH SPACES

J. E. Yukich*
Institut de Recherche Mathématique Avancée
Université Louis Pasteur
7 rue René Descartes
67084 Strasbourg, France

§1. INTRODUCTION

The intent of this article is to study the relationship between (i) the central limit theorem in Banach spaces, (ii) the Donsker property for unbounded classes of functions, especially subsets of the Banach dual, and (iii) metric entropy with L^p bracketing, $p \geq 1$.

Before exposing the main results, let us first set in place the framework for empirical processes, the setting of this paper. Throughout, take $(A, \mathcal{A}, P)$ to be a probability space and x_i, $i \geq 1$, the coordinates for the countable product $(A^\infty, \mathcal{A}^\infty, P^\infty)$ of copies of $(A, \mathcal{A}, P)$. The nth empirical measure for P is defined as

$$P_n(B) = n^{-1} \sum_{j=1}^{n} 1_{\{x_j \in B\}}, \quad B \in \mathcal{A}.$$

Given a class $\mathcal{F} \subset L^2(A, \mathcal{A}, P)$ of real-valued functions with envelope $F_{\mathcal{F}}(x) = \sup_{f \in \mathcal{F}} |f(x)|$, let $S := \ell^\infty(\mathcal{F})$ be the space of all bounded real-valued functions on $\mathcal{F}$; equip S with the sup norm, i.e., for $s \in S$ let

$$\| s \| := \sup_{f \in \mathcal{F}} |s(f)|.$$

We note that $(S, \| \ \|)$ is a Banach space, non-separable for $\mathcal{F}$ infinite.

When $F_{\mathcal{F}}$ is finite P a.e. the function-indexed empirical process

$$\nu_n(f)(\omega) := n^{1/2} \int f(dP_n - dP)(\omega), \quad \omega \in A^\infty, \quad f \in \mathcal{F},$$

is a random vector $\hat{v}_n$ with values in S. Recall [7,8] that $\mathcal{F}$ is a P-Donsker class if $\hat{v}_n$ converges in law in S to a Gaussian process G_P, indexed by $\mathcal{F}$, with a.s. bounded, ρ_P-uniformly continuous sample paths (abbreviated BUC). Here,

$$\rho_P^2(f,g) := \int (f-g)^2 \, dP - \left(\int (f-g) dP \right)^2.$$

*Current address: Lehigh University, Math., Bethlehem, Penn. 18015-USA

G_P necessarily has mean 0 and the same covariance as $\hat{v}_n$:

$$\operatorname{cov}(G_P(f), G_P(g)) = \int fg \, dP - \int f \, dP \int g \, dP .$$

Clearly, a necessary condition for F to be P-Donsker is that F be G_PBUC , i.e., that the limiting Gaussian process G_P can be chosen to be BUC.

Define the mapping $h: A \to S$ by setting

$$h(x)(f) = f(x) - \int f \, dP$$

for each $x \in A$. Defining the random variables $X_j = h(x_j)$, Dudley has shown [6] that if F is G_PBUC then the P-Donsker property for F means that X_1 satisfies the central limit theorem in $(S, \| \ \|)$, abbreviated $X_1 \in$ CLT . (Recall that if X is a random variable with values in a Banach space B and if $(X_n)_{n \in \mathbb{N}}$ is a sequence of independent copies of X , then X satisfies the CLT if

$$L\left(\sum_{i=1}^{n} X_i/\sqrt{n}\right)$$

converges weakly to a Radon measure on B . See [12,16] for an exposition of the CLT in Banach spaces.) In this way we see that central limit theorems for the empirical process $\nu_n(f)$, $f \in F$, can be viewed as central limit theorems with respect to the norm $\| \ \|$ on the Banach space S .

Throughout this paper let $(B, \| \ \|)$ denote a not necessarily separable Banach space , B^* the dual space, and B_1^* the unit ball of B^* . Now a subset H of B_1^* is called a norming subset (see [7]) for B if and only if $\|x\| = \sup_{h \in H} |h(x)|$ for all $x \in B$. Clearly, by the Hahn-Banach theorem, B_1^* is always a norming subset, a fact which we shall exploit in the second section.

In this way, limit theorems in B can be viewed as limit theorems for empirical measures on B , uniformly over a class of functions, such as B_1^* , since for $f \in B^*$ and $x(1),\ldots,x(n) \in B$

$$(\delta_{x(1)}+\ldots+\delta_{x(n)})(f) = f(x(1)+\ldots+x(n)) .$$

In particular, we note that X (with $L(X) = P$) satisfies the CLT in B if and only if B_1^* is P-Donsker.

The relationship between limit theorems for empirical processes and limit theorems for Banach space valued random variables has been studied by several authors. For example, Dudley has shown [7] that the Jain-Marcus CLT for $C(S)$-valued random variables [13] is in fact a consequence of Pollard's CLT [17] for empirical processes. Dudley has also shown [7] that the Fortet-Mourier strong law of large numbers [15] in separable Banach spaces is a consequence of the DeHardt-Dudley law of large numbers for a class of functions [3,7]; we will return to this implication later on and will also provide a short and simpler approach.

Limit theorems for $\nu_n(f)$ often involve a metric entropy condition on the index set F. In this article we will consider metric entropy with L^p bracketing, defined as follows [7]:

<u>Definition 1</u>. Given $f,g : A \to \mathbb{R}$, define the bracket $[f,g] := \{h:A \to \mathbb{R}$ such that $f(x) \leq h(x) \leq g(x)\ \forall x \in A\}$ and $\emptyset$ otherwise. Let $p \geq 1$, $\varepsilon > 0$ and $F \subset L^p(A,\mathcal{A},P)$. Let $N^{(p)}_{[\ ]} := N^{(p)}_{[\ ]}(\varepsilon,F,P) := \min\{m: \exists\ [f_1^-,f_1^+],\ldots$ $[f_m^-,f_m^+]$ such that $\forall f \in F\ \exists\ 1 \leq j \leq m$ such that $f \in [f_j^-,f_j^+]$ and $\int (f_j^+ - f_j^-)^p\, dP < \varepsilon^p\}$. Now $\log N^{(p)}_{[\ ]}(\varepsilon)$ is referred to as metric entropy with L^p bracketing.

We recall that metric entropy with L^1 bracketing has been especially useful in the study of empirical processes; see e.g. the works of DeHardt [3], Dudley [4,7], Dudley and Philipp [8], Alexander [1], Borisov [2], Giné and Zinn [11], Pyke [18], Kolčinskii [14], and Yukich [20,22, 23]. See also the recent monograph by Gaenssler [10]. In particular $N^{(1)}_{[\ ]}$ has been used to describe the weak convergence of $\nu_n(f)$ [4,7,8]. One of the main ideas running through this article is that L^1 bracketing is not a natural condition for describing weak convergence; L^2 bracketing is a more natural choice. On the other hand L^1 bracketing is a more natural condition for describing laws of large numbers. These points will be discussed in the following sections.

Finally, we will also use metric entropy without bracketing which has enjoyed wider use, and which is defined as follows:

<u>Definition 2</u>. Let $N^{(p)}(\varepsilon,F,P) := \min\{m: \exists\ f_1,\ldots,f_m$ such that for all $f \in F\ \exists\ 1 \leq j \leq m$ with $\int (f-f_j)^p dP < \varepsilon^p\}$. Now $\log N^{(p)}(\varepsilon,F,P)$ is referred to as the metric entropy of F.

Having set in place the framework for this paper, we now proceed to the main results, a summary of which was announced in [21]. Section two discusses the relationship between (i) and (iii) of the first paragraph, section three studies the relationship between (ii) and (iii), and section four considers metric entropy with L^2 bracketing.

§2. METRIC ENTROPY AND THE CLT FOR THE UNIT BALL OF THE BANACH DUAL

Our starting point is the following result of Dudley [7], which adds to Mourier's well known law of large numbers [15].

Theorem 1. (cf. section 6.1 of [7]) Let $X, X_1, \ldots$ be a sequence of i.i.d. random variables with values in a separable Banach space $(B, \|\ \|)$. The following are equivalent:

(i) X satisfies the strong law of large numbers (SLLN),

(ii) $E\,\|X\| < \infty$, and

(iii) $N^{(1)}_{[\ ]}(\varepsilon, B_1^*, P) < \infty \quad \forall \varepsilon > 0$.

It is reasonable to ask whether analogous results hold for those X satisfying the CLT. The following theorems answer this query in the affirmative, showing that the study of the CLT in separable Hilbert spaces H is conveniently studied through the use of $N^{(2)}_{[\ ]}(\varepsilon, H_1^*, P)$, where H_1^* is the unit ball of the Hilbert dual.

In fact, in the same way that the equivalence of

$$X \in \text{CLT} \quad \text{and} \quad E\{X\} = 0\ ,\ E\|X\|^2 < \infty$$

characterizes, modulo an isomorphism, separable Hilbert spaces, the following results show that the same is true for the equivalence of the conditions

$$B_1^* \text{ is P-Donsker and } \inf_{\varepsilon > 0} N^{(2)}_{[\ ]}(\varepsilon, B_1^*, P) < \infty\ ,\ P \text{ centered.}$$

This is contained in the following results.

Theorem 2. Let X be a random variable with values in $(B, \|\ \|)$, which need not be separable; $L(X) = P$. Then for all $p \geq 1$ the following are equivalent:

(i) $E\|X\|^p < \infty$, and

(ii) $\inf_{\varepsilon>0} N_{[\ ]}^{(p)} (\varepsilon, B_1^*, P) < \infty$.

If P is tight then the following are equivalent:

(iii) $E\|X\|^p < \infty$, and

(iv) $N_{[\ ]}^{(p)} (\varepsilon, B_1^*, P) < \infty \quad \forall \varepsilon > 0$.

Remarks

(1) Using the equivalence of (iii) and (iv) with $p = 1$ we directly obtain the double implication (ii)$\Leftrightarrow$(iii) of Theorem 1 without using the SLLN property of X.

(2) As is well known [12,16] the moment condition $E\|X\|^2 < \infty$ is in general neither necessary nor sufficient for $X \in CLT$; it may also be easily seen that the entropy condition $\inf_{\varepsilon>0} N_{[\ ]}^{(p)} (\varepsilon, F, P) < \infty$ is neither necessary nor sufficient for F to be P-Donsker. The interest of Theorem 2 stems from the fact that when F is B_1^* these are in fact equivalent conditions and they consequently share the same properties.

The next result is essentially a consequence of Theorem 2.

Theorem 3. Let X be a centered random variable with values in a Banach space $(B, \|\ \|)$; $L(X) = P$. Consider the following properties of P:

(i) $X \in CLT$,

(ii) B_1^* is a P-Donsker class of functions,

(iii) $E\|X\|^2 < \infty$,

(iv) $N_{[\ ]}^{(2)} (\varepsilon, B_1^*, P) < \infty \quad \forall \varepsilon > 0$,

(v) $\inf_{\varepsilon>0} N_{[\ ]}^{(2)} (\varepsilon, B_1^*, P) < \infty$.

We have:

(a) If P is tight and B is a Hilbert space then (i)-(v) are all equivalent.

(b) If B is separable, then the equivalence of (i), (ii), (iii), and (v) is equivalent to the fact that B is isomorphic to a Hilbert space.

For the proof, we need only observe that when B is a separable Hilbert space, the equivalence of (i), (ii) and (iii) stems from the introductory remarks.

For separable type 2 Banach spaces B we have sharp metric entropy conditions insuring the P-Donsker property for B_1^*.

Theorem 4. Let $(B, \|\ \|)$ be a separable type 2 Banach space and P a centered law. If

$$\inf_{\varepsilon>0} N_{[\ ]}^{(2)}(\varepsilon, B_1^*, P) < \infty,$$

then B_1^* is a P-Donsker class. Conversely, if P is a law on any separable Banach space B and if B_1^* is P-Donsker, then for all $p < 2$,

$$N_{[\ ]}^{(p)}(\varepsilon, B_1^*, P) < \infty \ \forall \varepsilon > 0 .$$

The proof of Theorem 4 follows at once from Theorem 2. For the first part we need only use implication (ii)$\Rightarrow$(i) with $p = 2$; for the second we use implication (iii)$\Rightarrow$(iv) with $p < 2$ together with the fact that if $X \in CLT$ then $E\|X\|^p < \infty \quad \forall p < 2$, see e.g. [16].

Proof of Theorem 2. We will first show (ii)$\Rightarrow$(i). Let $\varepsilon > 0$ and $m := m(\varepsilon) = N_{[\ ]}^{(p)}(\varepsilon, B_1^*, P)$. By definition there is a collection of brackets $[f_j^-, f_j^+]$, $j = 1, \ldots, m$, such that given any $f \in B_1^*$ $\exists\ 1 \le k \le m$ with

$$f_k^-(x) \le f(x) \le f_k^+(x) \qquad \forall x \in B$$

and

$$\int (f_k^+(x) - f_k^-(x))^p dP(x)) < \varepsilon^p .$$

Now B_1^* is a norming subset of B^* and thus for fixed $x \in B$ we see that $f(x)$ runs over all of the values between $-\|x\|$ and $\|x\|$ as f runs over B_1^*. This implies that for any fixed x, the sum of the differences of the brackets $\sum_{j=1}^{m} (f_j^+(x) - f_j^-(x))$ must be at least as large as $2\|x\|$. Indeed, suppose this were not the case. Then $\exists\ x_0 \in B$ such that

$$\sum_{j=1}^{m} f_j^+(x_0) - f_j^-(x_0) < 2\|x_0\| .$$

Thus there exists an $-\|x_0\| < \alpha < \|x_0\|$ and a $\delta > 0$ such that the open interval $(\alpha-\delta, \alpha+\delta)$ is disjoint from the union of closed intervals $\bigcup_{j=1}^{m} [f_j^-(x_0), f_j^+(x_0)]$. Moreover, by the Hahn-Banach theorem, there exists an $f \in B_1^*$ such that $f(x_0) = \alpha$. This leads to a contradiction since there exists $1 \leq j \leq m$ such that

$$f_j^-(x) \leq f(x) \leq f_j^+(x) \quad \forall x \in B .$$

Thus, we have shown that

$$\|X\| \leq \sum_{j=1}^{m} f_j^+(X) - f_j^-(X) \quad \text{a.s.}$$

For all $p \geq 1$ there exists a constant $C := C(p,m)$ such

$$\|X\|^p \leq C \sum_{j=1}^{m} ((f_j^+ - f_j^-)(X))^p \quad \text{a.s.}$$

Integrating this with respect to P and using the definition of $N_{[\ ]}^{(p)}(\varepsilon, B_1^*, P)$ shows that

$$E\|X\|^p \leq C \cdot m \cdot \varepsilon^p < \infty ,$$

completing (ii)$\Rightarrow$(i).

To show (i)$\Rightarrow$(ii) it suffices to consider the single bracket defined by $f_1^-(x) = -\|x\|$ and $f_1^+(x) = \|x\|$ and to take $\varepsilon = 2\int \|x\|^p \, dP(x)$.

If P is a tight measure then we may show the stronger implication (iii)$\Rightarrow$(iv); our proof is inspired by the proof of Proposition 6.1.7 of [7]. Let $p \geq 1$ be fixed and note that for all $\varepsilon > 0$ there is a compact set $K \subseteq B$ such that

$$\int_{B \setminus K} \|x\|^p \, dP < \varepsilon^p / 4 .$$

The elements of B_1^*, restricted to K, form a uniformly bounded equicontinuous family and hence this family is totally bounded for the sup norm $\| \|_K$ on K by the Arzelà-Ascoli theorem. Take $f_1,\dots,f_m \in B_1^*$, $m<\infty$, such that $\forall f \in B_1^*$

$$\|f-f_j\|_K < \varepsilon/4$$

for some j. Let $g_j = f_j - \varepsilon/4$ on K, $g_j(x) = -\|x\|$ for $x \notin K$, $g_{j+m} = f_j + \varepsilon/4$ on K, $g_{j+m}(x) = \|x\|$ for $x \notin K$, for all $j = 1,\dots,m$. Then for every $f \in B_1^*$, if $\|f-f_j\|_K < \varepsilon/4$, then $g_j \le f \le g_{j+m}$ and

$$\int |g_{j+m} - g_j|^p \le (2\varepsilon/4)^p + 2\int_{B\setminus K} \|x\|^p \, dP$$

$$\le (\varepsilon/2)^p + \varepsilon^p/2 \le \varepsilon^p .$$

Thus, $N_{[\,]}^{(p)}(\varepsilon, B_1^*, P) \le 2m < \infty$, proving (iii)$\Rightarrow$(iv) and completing the proof of Theorem 2. Q.E.D.

§3. THE P-DONSKER PROPERTY AND METRIC ENTROPY

The results of the above section show that $N_{[\,]}^{(2)}$ is often an appropriate tool for describing the P-Donsker property for B_1^*; the above results also suggest that $N_{[\,]}^{(p)}(\varepsilon, F, P)$, $p \neq 2$, will not in general be a satisfactory tool to describe the P-Donsker property for unbounded classes $F \subseteq L^p(A, A, P)$.

This is indeed the case: the following propositions show even in the presence of an envelope condition on F, that $N_{[\,]}^{(p)}(\varepsilon, F, P)$, $p \neq 2$, can not possibly give sharp results. This may be seen by considering suitably chosen subsets of the dual to $(\ell^\infty, \| \|_\infty)$, the Banach space of bounded functions on $\mathbb{N}^+$ equipped with the sup norm. More precisely, these subsets are classes of functions of the form

$$F_{\alpha,\beta} := \{\sum_j a_j s_j 1_{A_j} : s_j = 0 \text{ or } 1\} ,$$

where A_j , $j \geq 1$, is a sequence of disjoint subsets of A with $p_j := P(A_j) = j^{-\beta}$ for some $\beta > 1$ and $a_j = j^{\alpha}$ for some $0 < \alpha < \beta - 1$. Since elements of $F_{\alpha,\beta}$ define measures on $\mathbb{N}^+$, we may view $F_{\alpha,\beta}$ as a subset of the dual to $(\ell^{\infty}, \| \|_{\infty})$.

We will consider the cases $p = 1$, $p > 2$ and $1 < p < 2$ in this order; much of what follows may be found in [19]. Our first proposition shows that a theorem of Dudley, which is recalled below, is far from being the "best possible".

Theorem (cf. Theorem 3.1 of [5]). Suppose that F has envelope $F_F \in L^p(A,A,P)$ for some $p > 2$. Suppose that there exists γ , $0 < \gamma < 1 - 2/p$ and $M < \infty$ such that

$$N_{[\ ]}^{(1)} (\varepsilon, F, P) \leq \exp(M\varepsilon^{-\gamma})$$

for ε small enough. Then F is a P-Donsker class.

Proposition 1. For all $2 < p < 4$ there exist P-Donsker classes F with $F_F \in L^p(A,A,P)$ and $N_{[\ ]}^{(1)} (\varepsilon, F, P) \geq 2^{\varepsilon^{-\gamma}}$ where γ is any number less than 1/2.

Proof. Fix $2 < p < 4$ and choose τ and $\alpha > 0$ such that $\alpha + \tau + 2 < p/(p-2)$ and $\alpha(p-2) < 1$. Let $\beta = 2\alpha + 2 + \tau$ and let $F := F_{\alpha,\beta}$ as above.

Since $\beta > p\alpha + 1$ it is easily verified that $F_F \in L^p$. Also, by Theorem 2.4 of [5], F is a P-Donsker class since $\beta > 2\alpha + 2$ implies $\sum_j a_j \sqrt{p_j} < \infty$. It remains to find a lower bound for $N_{[\ ]}^{(1)} (\varepsilon, F, P)$. If the individual terms $a_j p_j$ are greater than or equal to ε for all $j \leq j_o = j_o(\varepsilon)$, then $N_{[\ ]}^{(1)} (\varepsilon, F, P) \geq 2^{j_o}$. Now $a_j p_j \geq \varepsilon$ if and only if $j^{\alpha-\beta} \geq \varepsilon$ and thus we may take $j_o \simeq \varepsilon^{-1/(\beta-\alpha)}$. Let $\gamma = 1/(\beta-\alpha) = 1/(\alpha+\tau+2)$. By taking α and τ small enough it is clear that γ may assume any value less than 1/2 . Q.E.D.

The next proposition shows that $N_{[\ ]}^{(1)}$ can not provide sharp results. In particular, if $2 < p < 3$ and if F is such that $F_F \in L^p$ and $N_{[\ ]}^{(1)} (\varepsilon, F, P) \simeq 2^{\varepsilon^{-\gamma}}$ for $(p-2)/(p-1) < \gamma < 1/2$, then F may or may not be a P-Donsker class of functions.

Proposition 2. For any $2 < p < 3$ and $\gamma > (p-2)/(p-1)$, there are classes F with envelope $F_F \in L^p$ and $N^{(1)}_{[\ ]}(\varepsilon, F, P) \le 2^{\varepsilon^{-\gamma}}$ which do not satisfy the P-Donsker property.

Proof. Let $2 < p < 3$ and find $\delta > 0$ such that $p < 3 - \delta$. Let $\alpha = (1-\delta)/(p-2)$, $\beta = 2\alpha + 2$, and consider the class $F = F_{\alpha,\beta}$. Now $F_F \in L^p$ but by Theorem 2.4 of [5], F is not P-Donsker.

Now if $j_0 = K\varepsilon^{-1/(\beta-\alpha-1)}$ for some suitably large constant K, then $\sum_{j \ge j_0} a_j p_j = \sum_{j \ge j_0} j^{\alpha-\beta} \le \varepsilon$ and thus $N^{(1)}_{[\ ]}(\varepsilon, F, P) \le 2 \cdot 2^{j_0}$ Setting $\gamma = 1/(\beta-\alpha-1) = (p-2)/(p-1-\delta)$ and letting $\delta \downarrow 0$ we see that $\gamma \downarrow (p-2)/(p-1)$, giving the desired result. Q.E.D.

We conclude this section by exploring the relationship between P-Donsker classes and $N^{(p)}_{[\ ]}(\varepsilon, F, P)$, $p > 1$. We consider the cases $p > 2$ and $1 < p < 2$ in this order. We first show that $N^{(p)}_{[\ ]}$, $p > 2$, will not in general furnish sharp results for the P-Donsker property.

Proposition 3. If $q > 2$ and $p < q/2 + 1$, then there are classes F with $F_F \in L^p$ and $N^{(q)}_{[\ ]}(\varepsilon, F, P) \simeq 2^{\varepsilon^{-\gamma}}$, $\gamma > q$, such that F may or may not be a P-Donsker class.

Proof. We first show that there are P-Donsker classes F with $F_F \in L^p$, $p < q/2 + 1$, such that $N^{(q)}_{[\ ]}(\varepsilon, F, P) \ge 2^{\varepsilon^{-\gamma}}$ for $\gamma > q$ arbitrarily large. Let $q > 2$ and $p < q/2 + 1$ be fixed, let $\delta > 0$, $\alpha = 2/q-2)$, $\beta = 2\alpha + 2 + \delta$ and show that if $F := F_{\alpha\,\beta}$ then $N^{(q)}_{[\ ]}(\varepsilon, F, P)$ becomes arbitrarily large for δ arbitrarily small. To see that this is actually so, note that if the individual terms $a_j^q p_j \ge \varepsilon^q$ for all $j \le j_0(\varepsilon)$, then $N^{(q)}_{[\ ]}(\varepsilon, F, P) \ge 2^{j_0}$. Now $a_j^q p_j \ge \varepsilon^q$ if and only if $j^{q\alpha-\beta} \ge \varepsilon^q$ if and only if $j_0 \simeq \varepsilon^{-q/(\beta-q\alpha)}$. If $\gamma = q/(\beta-q\alpha)$ then $N^{(q)}_{[\ ]}(\varepsilon, F, P) \ge 2^{\varepsilon^{-\gamma}}$; moreover $\gamma = q/s \uparrow \infty$ as $\delta \downarrow 0$. Finally, we note that F is P-Donsker and $F_F \in L^p$, since $\beta > p\alpha + 1$.

On the other hand, if $N^{(q)}_{[\ ]}(\varepsilon, F, P) \le 2^{\varepsilon^{-\gamma}}$ for $\gamma > q > 2$, then F may not be P-Donsker, even if F is of the form $F_{\alpha,\beta}$. Indeed, let $0 < \alpha < 1/(q-2)$, $\beta = 2\alpha + 2$, and $F = F_{\alpha,\beta}$. If there is a $j_o = j_o(\varepsilon)$ such that

$$\int (\sum_{j \geq j_o} a_j 1_{A_j})^q \, dP \leq \varepsilon^q ,$$

then $N^{(q)}_{[\ ]}(\varepsilon, F, P) \leq 2 \cdot 2^{j_o}$. Since the A_j are disjoint, this inequality is satisfied whenever $\sum_{j > j_o} j^{q\alpha-\beta} \leq \varepsilon^q$, i.e., whenever $j_o^{q\alpha-\beta+1} \simeq \varepsilon^q$. Thus, if $\gamma = q/(\beta - q\alpha - 1)$, then $N^{(q)}_{[\ ]}(\varepsilon, F, P) \leq 2^{\varepsilon^{-\gamma}}$. Letting $\alpha \downarrow 0$ we obtain $\gamma \downarrow q$, completing the proof. Q.E.D.

Finally, we consider $N^{(q)}_{[\ ]}(\varepsilon, F, P)$ with $1 < q < 2$.

Proposition 4. If $1 < q < 2$ is fixed, $2 < p < 4-q$, and $q(p-2)/(p-q) < \gamma < q/2$, then there exist classes F with $F_F \in L^p$, $N^{(q)}_{[\ ]}(\varepsilon, F, P) \leq 2^{\varepsilon^{-\gamma}}$, such that F may or may not be P-Donsker.

Proof. Let $1 < q < 2$ and $2 < p < 4-q$ be fixed. Let $\alpha < 1/(p-2)$ be fixed and let $\beta := 2\alpha + 2$. Then $F := F_{\alpha,\beta}$ is not P-Donsker and $F_F \in L^p$. The computations of the second half of the proof of Proposition 3 show that if $\gamma \simeq q/(\beta - q\alpha - 1) = q/((2-q)\alpha + 1)$, then $N^{(q)}_{[\ ]}(\varepsilon, F, P) \leq 2^{\varepsilon^{-\gamma}}$. Letting $\alpha \uparrow 1/(p-2)$ shows that $\gamma \downarrow q(p-2)/(p-q)$, as desired.

In a similar way, the first half of the proof of Proposition 3 shows that F may not be P-Donsker. Let δ, α, β be as in the proof of Proposition 3. As $\alpha \downarrow 0$ and $\delta \downarrow 0$ we see that $\gamma \uparrow q/2$, as desired. Q.E.D.

§4. METRIC ENTROPY WITH L^2 BRACKETING

Dudley (c.f. Theorem 6.2.1 of [7]) has shown that if F is a uniformly bounded class of functions on $(A, \mathcal{A}, P)$ with

$$\int_0 \sqrt{\log N_{[\ ]}^{(1)}\ (\varepsilon^2, F, P)}\ d\varepsilon < \infty \ , \tag{1}$$

then F is a P-Donsker class. The above sections suggest that $N_{[\ ]}^{(2)}\ (\varepsilon, F, P)$ is a more natural choice for the description of the P-Donsker property; we are thus lead to the following

Conjecture 1. Let $F \subset L^2(A, A, P)$ with $F_F \in L^2(A, A, P)$; suppose that

$$\int_0 \sqrt{\log N_{[\ ]}^{(2)}\ (\varepsilon, F, P)}\ d\varepsilon < \infty \ . \tag{2}$$

Then F is a P-Donsker class of functions.

Note that this conjecture should be compared with Pollard's central limit theorem [17] where $N_{[\ ]}^{(2)}$ is replaced by a random entropy. Also, for uniformly bounded F, (2) is generally weaker than (1).

Now the following theorems support the conjecture and show that condition (2) is actually necessary in some cases. Recall that $N^{(2)}$ is given as in Definition 2.

Theorem 5. Let P be a probability measure on $\mathbb{R}$ with a density $f(x)$ such that $(f(x)+f(-x))$ is decreasing for x large; let H be the class of functions $\{x \to e^{itx} : |t| \leq 1\}$. The following are equivalent:

(i) $\int_0 \sqrt{\log N_{[\ ]}^{(1)}(\varepsilon^2, H, P)}\ d\varepsilon < \infty$,

(ii) $\int_0 \sqrt{\log N_{[\ ]}^{(2)}(\varepsilon, H, P)}\ d\varepsilon < \infty$,

(iii) $\int_0 \sqrt{\log N^{(2)}(\varepsilon, H, P)}\ d\varepsilon < \infty$, and

(iv) H is P-Donsker .

The proof of Theorem 5 closely parallels the proof of Theorem 1 of [22] where the equivalence of (i), (ii), and (iv) is demonstrated; we do not provide the details here.

Our final theorem should be regarded as a generalization of the Borisov-Dudley-Durst theorem [2,7,9] which characterizes when the class of all subsets of $\mathbb{N}^+$ is P-Donsker.

Theorem 6. Let f_m , $m \geq 1$, be a sequence of positive functions on $\mathbb{R}$ with disjoint support; let $\|f_m\|_2 = (\int f_m^2 dP)^{1/2}$ and suppose $\sum_m \|f_m\|_2^2 < \infty$. Let

$$F := \{ \sum_{m \in A} f_m : A \subseteq 2^{\mathbb{N}^+} \} .$$

The following are equivalent:

(i) $\sum_m \|f_m\|_2 < \infty$,

(ii) $\int \sqrt{\log N_{[\ ]}^{(2)}(\varepsilon, F, P)}\, d\varepsilon < \infty$, and

(iii) F is P-Donsker.

Proof. Observe that standard Gaussian processes techniques give (i)$\Leftrightarrow$(iii), see [19] for details.

Let us prove (i)$\Leftrightarrow$(ii); this will essentially be a consequence of the Borisov-Dudley-Durst theorem.

Take $\sum_m \|f_m\|_2^2 = M < \infty$ and let Q be a probability measure on $\mathbb{N}^+$ defined by

$$Q(\{m\}) = \|f_m\|_2^2 / M .$$

Now we claim that

$$N_{[\ ]}^{(2)}(\varepsilon, F, P) = N_{[\ ]}^{(1)}(\varepsilon^2/M,\ 2^{\mathbb{N}^+}, Q) \tag{3}$$

where, as usual, the metric entropy for a class of sets is taken to be the metric entropy for the class of respective indicator functions.

To show (3) we show that every bracket for the class F corresponds to a bracket for $2^{\mathbb{N}^+}$ and vice versa. By minimality of $N_{[\ ]}^{(2)}(\varepsilon, F, P)$, a bracket $[h_i^-, h_i^+]$ for F will necessarily have the form

$$h_i^- = \sum_{m \in A_i^-} f_m \quad \text{and} \quad h_i^+ = \sum_{m \in A_i^+} f_m$$

for some sets A_i^- and A_i^+ , $A_i^- \subseteq A_i^+$, belonging to $2^{\mathbb{N}^+}$. Now observe that

$$\int (h_i^+ - h_i^-)^2 dP = \int \Big(\sum_{m \in A_i^+ \setminus A_i^-} f_m \Big)^2 dP$$

$$= \int \sum_{m \in A_i^+ \setminus A_i^-} f_m^2 dP$$

$$= \sum_{m \in A_i^+ \setminus A_i^-} \|f_m\|_2^2$$

$$= M \cdot \int (1_{A_i^+} - 1_{A_i^-}) dQ .$$

Thus, if $\{\int (h_i^+ - h_i^-)^2 dP\}^{1/2} < \varepsilon$, then $\int (1_{A_i^+} - 1_{A_i^-}) dQ < \varepsilon^2/M$, showing that every $L^2(P)$ ε-bracket for F is an $L^1(Q)$ ε^2/M-bracket for $2^{\mathbb{N}+}$. This shows $N_{[\ ]}^{(2)}(\varepsilon, F, P) \geq N_{[\ ]}^{(1)}(\varepsilon^2/M, 2^{\mathbb{N}+}, Q)$; to show the reverse inequality we just need to trace these steps backwards.

By the Borisov-Dudley-Durst theorem, we have

$$\int_0 \sqrt{\log N_{[\ ]}^{(1)}(\varepsilon^2, 2^{\mathbb{N}+}, Q)}\, d\varepsilon < \infty \quad \Leftrightarrow \quad \sum_m \|f_m\|_2 < \infty . \tag{4}$$

Combine (3) and (4) to obtain (i)$\Leftrightarrow$(ii); this completes the proof of Theorem 6. Q.E.D.

§5. FINAL REMARKS

Having shown the efficacy of metric entropy with L^2 bracketing, much still remains to be done. Apart from resolving the validity of the conjecture, it remains to be seen whether $N_{[\ ]}^{(2)}$ proves useful in studying the bounded and compact laws of the iterated logarithm (LIL). Based on the results of [20,23] it seems reasonable to make the following

<u>Conjecture 2.</u> Let $\mathcal{F} \subset L^2(A,\mathcal{A},P)$ with $F_{\mathcal{F}} \in L^2(A,\mathcal{A},P)$; suppose that

$$\int_0 \sqrt{\frac{\log N^{(2)}_{[\ ]}(\varepsilon,\mathcal{F},P)}{\log\log\log N^{(2)}_{[\ ]}(\varepsilon,\mathcal{F},P)}}\, d\varepsilon < \infty .$$

Then

$$\overline{\lim_{n\to\infty}} \sup_{f\in\mathcal{F}} \frac{|\nu_n(f)|}{\sqrt{\log\log n}} < \infty \quad \text{a.s.}$$

Were it true, this conjecture would establish sufficient metric entropy conditions for the bounded LIL weaker than those provided by [8].

Whether $N^{(2)}_{[\ ]}$ provides general rates of convergence for $\nu_n(f)$ more refined than those previously established by L^1 bracketing techniques (see e.g. [1,8,14,23]) is also an open question.

<u>Note added in proof</u>: After the writing of this paper I was informed that Dr. M. Ossiander has obtained a proof of the conjecture announced in section 4; at the time of writing, however, I have not yet seen Ossiander's proof.

REFERENCES

1. Alexander, K. S. (1984). Rates of growth and sample moduli for weighted empirical processes indexed by sets, preprint.

2. Borisov, I. S. (1983). Problem of accuracy of approximation in the central limit theorem for empirical measures, Siberskij Matematicheskij Zhurnal, v. 24, no. 6, pp. 14-25 = Siberian Mathematical Journal, July issue, 1984, pp. 833-843.

3. DeHardt, J. (1971). Generalization of the Glivenko-Cantelli theorem, Ann. Math. Statist., 42, pp. 2050-2055.

4. Dudley, R. M. (1978). Central limit theorems for empirical measures, Ann. Prob. 6, pp. 899-929; correction 7, (1979) pp. 909-911.

5. Dudley, R. M. (1981). Donsker classes of functions, Statistics and Related Topics (Proc. Symp. Ottawa 1980), pp. 341-352, New York, Amsterdam: North Holland.

6. Dudley, R. M. (1984). An extended Wichura theorem, definitions of Dansker class, and weighted empirical distributions, preprint.

7. Dudley, R. M. (1984). A course on empirical processes, Lecture Notes in Mathematics, no. 1097.

8. Dudley, R. M. and Walter Philipp (1983). Invariance principles for sums of Banach space valued random elements and empirical processes, Z. Wahrschein. verw. Geb., 62, pp. 509-552.

9. Durst, Mark and R. M. Dudley (1981). Empirical processes, Vapnik-Chervonenkis classes and Poisson processes, Prob. Math. Statist. (Wrocław), 1, no. 2, pp. 109-115.

10. Gaenssler, P. (1983). Empirical processes. Institute of Mathematical Statistics Lecture Notes - Monograph Series 3.

11. Giné, E. and J. Zinn (1984). Some limit theorems for empirical processes, Ann. Prob., 12, pp. 929-989.

12. Jain, N. (1977). Central limit theorem and related questions in Banach space, Proceedings of Symposia in Pure Mathematics, vol. 31, pp. 55-65.

13. Jain, N. and M. B. Marcus (1975). Central limit theorems for C(S) valued random variables, J. Functional Analysis, 19, pp. 216-231.

14. Kolčinskii, V. I. (1981). On the central limit theorem for empirical measures, Theor. Probability Math. Statist., 24, pp. 71-82 = Teor. Verojatnost. i. Mat. Statist., 24, pp. 63-75.

15. Mourier, Edith (1951). Lois des grands nombres dans un espace de Banach, C. R. Acad. Sci. Paris, 232, pp. 923-925.

16. Pisier, G. (1975). Le theórème de la limite centrale et la loi du logarithme itéré dans les espaces de Banach, suite et fin, Sém. Maurey-Schwartz 1975-76, Exposé IV, Ecole Polytechnique, Palaiseau.

17. Pollard, David B. (1982). A central limit theorem for empirical processes, J. Australian Math. Soc., Ser. A. 33, pp. 235-248.

18. Pyke, Ronald (1984). Asymptotic results for empirical and partial-sum processes: A review, The Canadian Journal of Statistics, 12, no. 4, pp. 241-264.

19. Yukich, J. E. (1982). Convergence of empirical probability measures, Ph.D. thesis (M.I.T.)

20. Yukich, J. E. (1984). The law of the iterated logarithm and the empirical characteristic function, preprint.

21. Yukich, J. E. (1985). Théorème limite central et entropie métrique dans les espaces de Banach, C. R. Acad. Sci. Paris, to appear.

22. Yukich. J. E. (1985). Weak convergence of the empirical characteristic function, Proc. of Amer. Math. Soc., to appear.

23. Yukich. J. E. (1985). Rates of convergence for classes of functions, preprint.